C0-AJX-615

The World of Microbes

The World of Microbes

Howard Gest

Science Tech Publishers, Inc.
Madison, Wisconsin

The Benjamin/Cummings Publishing Company, Inc.
Menlo Park, California • Reading, Massachusetts • Don Mills, Ontario

Printed in the United States of America.

10 9 8 7 6 5 4 3 2 1

Library of Congress Cataloguing-in-Publication Data
Gest, Howard
The world of microbes

Bibliography: p.
Includes index.
1. Microbiology I. Title. [DNLM: 1. Microbiology.
QW 4 G393w]
QR41.2.G47 1987 576 87-12664
ISBN 0-910239-10-X

Production and editorial services: Science Tech Publishers
Artist: Edward B. Phillips
Design: Katherine M. Brock
Cover art reproduced from the end papers of *Microbes of Merit*, by Otto Rahn, 1945. Jaques Cattell Press, Lancaster, PA.

This book is a copublication of Science Tech Publishers and The Benjamin/Cummings Publishing Company. Sales are handled as follows:

Trade and individual sales: Science Tech Publishers,
701 Ridge Street, Madison, Wisconsin 53705
ISBN 0-910239-10-X

Text sales and adoptions: The Benjamin/Cummings Publishing Company, Inc.,
2727 Sand Hill Road, Menlo Park, California 94025
ISBN 0-8053-2820-3

Where the telescope ends, the microscope begins. Which of the two has the grander view?

Victor Hugo, *Les Misérables* (1862)

About the Author

Howard Gest holds the title of Distinguished Professor of Microbiology at Indiana University, Bloomington. Born in London, England in 1921, Professor Gest has had a distinguished scientific career that has extended over almost 40 years. He received his Bachelor of Arts degree in bacteriology from the University of California, Los Angeles in 1942 and his Ph.D. in microbiology from Washington University, St. Louis in 1949. During World War II he was associated with the Manhattan Project on the development of nuclear energy. He has been on the faculty of Case Western Reserve University, Washington University, and Indiana University and has been a visiting researcher at the California Institute of Technology, Dartmouth Medical School, Stanford University, Oxford University, Tokyo University, and U.C.L.A. He has twice been a Guggenheim Fellow, and has served on a number of advisory committees of the United States government. He has done research on numerous aspects of microbial physiology and biochemistry, especially with photosynthetic bacteria, and has studied such processes as nitrogen fixation, hydrogen metabolism, regulation of amino acid biosynthesis, and various bioenergetic mechanisms. During his second Guggenheim Fellowship he studied problems of biochemical evolution at U.C.L.A. as a member of the Precambrian Paleobiology Research Group. As one of his collegues has written: "Professor Gest's contributions cannot be measured solely on the basis of his research. There is also an intangible aspect. He is a kind and thoughtful man, always supportive of younger faculty and graduate students, and a patient teacher . . . He truly loves science and has imparted this enthusiasm to a younger generation of scientists. He has thus been a successful teacher as well as a productive researcher."

Preface

Microbiology is concerned with the properties of forms of life not visible to the naked eye, microbes. They are found virtually everywhere and they affect our lives in numerous ways. During the last 100 years our knowledge of both the beneficial and the deleterious effects of microbes on animals and plants has increased greatly, and today microbiology has become one of the most exciting disciplines of modern science.

Two major developments led to the emergence of microbiology as a dynamic force in the current enterprises of science and technology. The first one is the realization that microbes provide unique experimental systems for studying the basic process of *all* forms of life. Secondly, the exploitation of microbes for biotechnological applications now appears to have almost unlimited horizons. In the decades ahead, these applications of microbiology will produce great changes in medicine, public health, agriculture, space exploration, environmental pollution, and industrial manufacturing.

Microbes have been continually recycling chemical elements on the Earth for hundreds of millions of years and if these changes did not occur, all life on our planet would soon come to an end. To understand these phenomena, and how microbes interact with plants and animals, it is essential to have an appreciation for some rudiments of chemistry. This poses problems in explaining microbiology simply and clearly to the lay person, the nonscientist, the business person, the entrepreneur, the technician, or humanist: in short, for the person who does not have some training in chemistry. Therefore, I have tried to write a "guidebook" that describes in a simplified but authoritative fashion the many interactions of microbes with higher forms of life and our environment, and the essence of microbial biotechnology.

The continual challenge of teaching a course in microbiology for nonscientists has led me to spend many hours in libraries and bookshops searching for a suitable text. Despite the fact that microbes are in the news almost every day and are currently the subject of numerous articles in the popular press, it remains difficult to find books that explain the facts and principles of modern microbiology in a way that can be easily understood by people who have had little formal scientific training, or even by scientists in nonbiological fields. Thus, until now my students have had to make do with books that are either not at an appropriate level, or are devoid of the excitement of science. I wrote this book in order to attempt to fill this gap, assuming that the reader has little, if any, formal knowledge of chemistry.

Year after year, I have been able to count on the regular appearance of news articles about microbes which could be used to add spice to my lectures. A timely example: while I was writing this preface, a newspaper article reported that scientists had been able to use biotechnology to create tobacco plants which glowed in the dark. These scientists transferred a gene from the firefly to the tobacco plant, using a bacterium as an intermediary. The "engineered" plants were said to "give off a low, soft glow in the dark." The purpose of the experiment was not, however, to produce luminescent plants (we obviously don't need tobacco plants that glow in the dark), but to develop a technique for transferring genes from one species to another and to study how these genes are turned on and off—"expressed"—in cells.

Since the end of World War II we have witnessed an avalanche of basic discoveries, many of which have been applied to improve agriculture, public health and industrial processes. It has been demonstrated repeatedly that scientific discoveries and technologies derived from these discoveries strongly influence the course of human history. An example of the historical context of scientific advances is provided by the British physicist John Ziman:*

> In all human affairs. . .there is a single dominant variable—time. To make sense of the present state of science, we need to know how it got like that: we cannot avoid an *historical* account. . . .To extrapolate into the future we must look backwards a little into the past.

My own study of the history of microbiology and biochemistry has convinced me that the major advances, in the form of dramatic leaps in insight, have usually been the product of either an individual researcher or of a few scientists obsessed with the desire to explain a basic biological phenomenon. Although in some fields of science the current *style* involves large teams of researchers, I believe that the truly innovative advances at the forefront of new knowledge will continue to be made by individuals or small groups of scientists. In this book, I discuss some of the pioneers of microbiology whose work changed the course of history. In reflecting on the dynamics of scientific discoveries, I have on many occasions recalled the following quotation which I first read when I was beginning my own college studies:

> There are ancient cathedrals which, apart from their consecrated purpose, inspire solemnity and awe. Even the curious visitor speaks of serious things, with hushed voice, and as each whisper reverberates through the vaulted nave, the returning echo seems to bear a message of mystery. The labor of generations of architects and artisans has been forgotten, the scaffolding erected for their toil has long since been removed, their mistakes have been erased, or have become hidden by the dust of centuries.

---

*Ziman, J.M., 1976, *The Force of Knowledge.* Cambridge University Press, Cambridge.

> Seeing only the perfection of the completed whole, we are impressed as by some superhuman agency. . . .
> Science has its cathedrals, built by the efforts of a few architects and of many workers.*

Microbiology has also had its own architects and its many artisans.

It is a pleasure to record my thanks to friends who aided in the development of this book. In particular, I am indebted to Michael Madigan (Southern Illinois University), Joel Mandelstam (University of Oxford), Barry Marrs (E. I. duPont de Nemours and Co.), and Robert Gherna (American Type Culture Collection) for many helpful suggestions. I am also grateful for comments from Philip Beer (Milford, Indiana) and from my colleagues at Indiana University—Lori Mangels, Peggy Beer-Romero, David White, Fredric Brewer, Jeffrey Favinger, and John Gallman. My greatest debt is to my wife for her unfailing encouragement, patience, and willingness to debate the merits of dozens of alternative titles for this attempt to explain microbes to a wide audience.

---

*Lewis, G.N. and Randall, M., 1923, *Thermodynamics and the Free Energy of Chemical Substances.* McGraw-Hill, New York.

Contents

	Preface	vii
1	Leeuwenhoek Discovers a New Galaxy of Organisms	1
2	The Microbial Kingdom Has Many Subjects	9
3	Some Microbes Prefer Life Without Air	13
4	Important Molecules in Microbes, Plants, and Animals	19
5	How Microbes are Isolated and Identified	33
6	The Care and Feeding of Microbes	45
7	Hardy Survivors in the Microbial Kingdom	53
8	Microbes and the Carbon Cycle	62
9	Bacteria That Produce and Use Methane	72
10	Microbes Recycle Nitrogen	82
11	Bacteria Spin the Sulfur Cycle	91
12	An Amazing Diversity of Lifestyles	95
13	Bioenergetics: "Energy Currency"	116
14	The Role of Vitamins	129
15	Microbes and Sewage Treatment	135
16	Plagues and the Origin of the Germ Theory of Disease	144
17	Three Giants of Infectious Disease Research: Pasteur, Koch, and Jenner	153
18	Mechanisms of Immunity	161
19	Viruses Confound Microbe Hunters	175
20	The Control of Microbial Disease	181

21 The Role of DNA and New Vistas in Microbial Technology 187
22 Coda: Microbes and Early Life on Earth 207
Appendix I. How Leeuwenhoek Estimated the Sizes of Microbes 212
Appendix II. Microbes in the American Type Culture Collection 215
Appendix III. Microbes in Early Science Fiction 217
Appendix IV. The Ingenious Use of Microbiology Under Adverse Conditions 222
Who's Who in This Book 228
Suggestions for Further Reading 232
Glossary 236
Credits and Acknowledgments 241
Index 243

The World of Microbes

1

Leeuwenhoek Discovers a New Galaxy of Organisms

Microbiology is the science that deals with microorganisms, which are also known as *microbes*. The word microbe was first used in 1878 to describe "extremely minute living beings." At that time, the term was chiefly applied to one major category of microbes, the bacteria. Before 1878, scientists—including Louis Pasteur—used a variety of terms rather loosely to label the very small organisms that interested them (for example: "animalcules," "infusoires," "germs"). It was not clear whether microbes belonged to the animal or plant kingdoms, or somewhere else. Nineteenth century investigators also did not fully realize that life on Earth as we know it could not exist without the activities of a large collection of microbes that are invisible to the naked eye. *Visible effects* of microbes on higher plants and animals, however, were commonplace and evident long before the existence of microbes was discovered in the seventeenth century. When the effects were deleterious—for example, in the form of infectious diseases—they were particularly obvious and were viewed as supernatural events or mysterious "spontaneous" phenomena. Rational explanation of infectious disease and other manifestations of microbial life had to await two developments: acceptance of the *concept* that "invisible microbes" existed and tangible *evidence* of their reality.

The first evidence that we are surrounded by multitudes of microbes was provided by observations made by Antoni van Leeuwenhoek with primitive microscopes in 1674. This historic discovery revealed not only the physical reality of living microbes, but also their diverse nature. These momentous advances, which are discussed in this chapter and in Chapter 2, illustrate one of the common characteristics of all waves of new discoveries in biological science, namely, the use of new or improved experimental techniques for making observations.

The English scientist Robert Hooke (1635–1702) has the distinction of contributing to the improvement of almost every important scientific instrument developed during the seventeenth century. His famous book *Micrographia* (1665) expounded numerous uses of the microscope for study of biological science, and his observations on cork led him to coin the word *cell* to describe the basic units of biological structure:

> I Took a good clear piece of Cork, and with a Pen-knife sharpen'd as keen as a Razor, I cut a piece of it off, and thereby left the surface of it exceeding smooth, then examining it very diligently with a *Microscope*, me thought I could perceive it to appear a little porous. . . .[I then] cut off from the former smooth surface an exceeding thin piece of it, and placing it on a black object Plate, because it was itself a white body, and casting the light on it with a deep *plano-convex Glass*, I could exceeding plainly perceive it to be all perforated and porous, much like a Honey-comb, but that the pores of it were not regular; yet it was not unlike a Honey-comb in these particulars.

With his microscope, which magnified about 25 times, Hooke saw similar textures in other kinds of plant tissues (tissues are aggregates of cells that are bound together to perform one or more functions).

After Hooke's time, microscopes were gradually improved,* and eventually in 1838 it was recognized that all plants and animals are composed of cells, a concept that was called the "cell theory." Different types of cells vary greatly in size. A single human nerve cell can be as long as 3 to 4 feet. An ostrich egg cell is usually the size of a small grapefruit, but most cells of animals and plants are in the range of 10 to 100 micrometers in diameter. To get your bearings in this microscopic world, consider that 1 micrometer is one-millionth of a meter. To put it another way, 10,000 micrometers corresponds to 1 centimeter (1 centimeter equals 0.39 inch). Despite the small sizes of typical plant and animal cells, they are extremely complicated, both with respect to details of internal structure and how the cell "machinery" works.

In 1674, nine years after Hooke first described cork cells, a Dutch shopkeeper discovered the existence of living cells even smaller than those of plants and animals. This remarkable event resulted from the insatiable curiosity and great skills of Antoni van Leeuwenhoek (1632–1723) who had little formal education and had never attended a university. As a draper, he dealt with cloth, ribbons, buttons, and the like. Careful drapers were in the habit of using a low-power magnifying glass to inspect the quality of cloth, and this was the starting point of Leeuwenhoek's unique scientific career. He had the ability to make small lenses—only about 1 millimeter in diameter—of superb quality. Each lens was embedded in a small metal sheet (about 1 × 2 inches), and the device was equipped with adjustable screws that could position a sample (for example, something contained in a very thin glass tube) near the lens (Figure 1). When held close to the eye and focused by adjusting the screws, these simple microscopes revealed to Leeuwenhoek clear images of very small objects, magnified as many as 300 times or more.

---

*Excellent collections of early and modern microscopes can be seen at the Armed Forces Medical Museum, Armed Forces School of Pathology, Washington, D.C.; at the Science Museum, London; and at the Museum of the History of Science, University of Oxford.

(a)

Figure 1 (a) Antoni van Leeuwenhoek. (b, *next page*) A replica of one of Leeuwenhoek's microscopes. The object to be viewed was placed on the pointed tip at the end of the screw. The lens is in the small circle just to the left of the screw tip.

Leeuwenhoek can be regarded as one of the great explorers of all time—indeed, he discovered a whole new world by examining an enormous range of natural samples. In the course of his studies, he described for the first time the sperm cells of animals, including humans, and he was also the first person to recognize that in the fertilization process, the sperm enters the egg cell. He provided the first accurate description of red blood cells. At a time when it was widely thought that maggots, fleas,

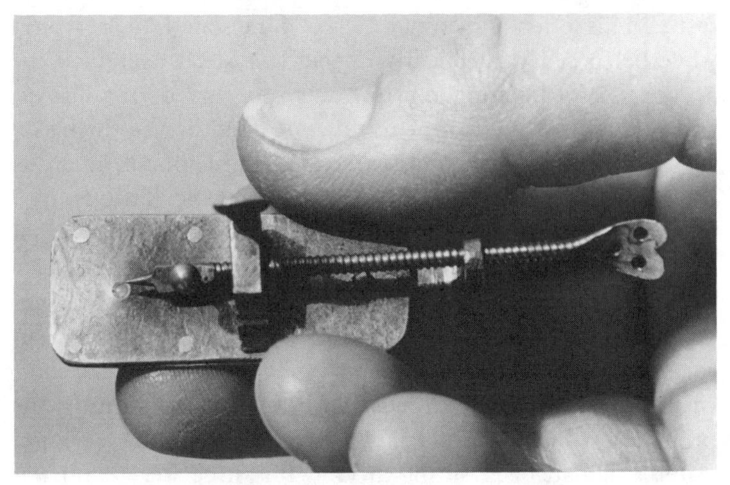

(b)

and the like were formed by "spontaneous generation," Leeuwenhoek showed that such creatures hatch from fertilized eggs. The list of "firsts" goes on and on. To measure objects in this new "invisible" world, Leeuwenhoek had to devise new reference standards, such as the diameters of a grain of coarse sand (870 micrometers), a hair from his beard (100 micrometers), and a human red blood cell (7.5 micrometers) (see Appendix I).

Without doubt, Leeuwenhoek's greatest contribution to biology was the discovery of microbes, the smallest forms of which are bacteria. He described his findings in minute detail in a series of letters sent to the Royal Society of London, and this collection is an outstanding classic work of biological research. His letters created a sensation, and some Fellows of the prestigious society found it hard to believe a number of his claims. Leeuwenhoek consequently felt obliged to have testimonials about the reliability of his observations sent to the Royal Society by Dutch ministers, physicians, and jurists!

His discovery of microbes is an interesting example of serendipity in research that began with an interest in the sense of taste. In his letter of October 19, 1674, he stated, "Last winter while being sickly and nearly unable to taste, I examined the appearance of my tongue, which was very furred, in a mirror, and judged that my loss of taste was caused by the thick skin on the tongue." This led him to examine little points on an ox tongue with his microscope, and he saw that the "little points" had "very fine pointed projections" that were composed of "very small globules." Obviously, he was observing the taste buds, but he continued to be mystified about why pepper, ginger, nutmeg, cloves, etc., have such potent tastes. So he performed many kitchen experiments which took the form of soaking or pounding the spices in water and examining the preparations. On April 24, 1676, he scrutinized some pepper water that had been sitting around for three weeks and was astonished to observe many very small organisms that he called "animalcules." His animalcules (or "little eels") were actually bacteria, which typically are only about 1 to 2 micrometers in diameter. He immediately looked for the animalcules in other places, for example, in the white matter that he found stuck to his teeth: "I have mixed it with clean rain water, in which there were no 'animalcules,' and I most always saw with great wonder that there were many very little animalcules, very prettily a-moving."* Naturally, he became interested in the mouths of other people. Here is an excerpt from Letter 39 to the Royal Society:

> While I was talking to an old man (who leads a sober life, and never drinks brandy or tobacco, and very seldom any wine), my eye fell upon his teeth, which were all coated over; so I asked him when he had last cleaned his mouth? And I got for answer that he'd never washed his mouth in all his life. So I took some spittle out of his mouth and examined it; but I could find in it

---

*For a detailed account of how Leeuwenhoek's interest in taste led to the discovery of bacteria, *see* Bardell, D., 1983, The roles of the sense of taste and clean teeth in the discovery of bacteria by Antoni van Leeuwenhoek. *Microbiological Reviews,* 47:121–126.

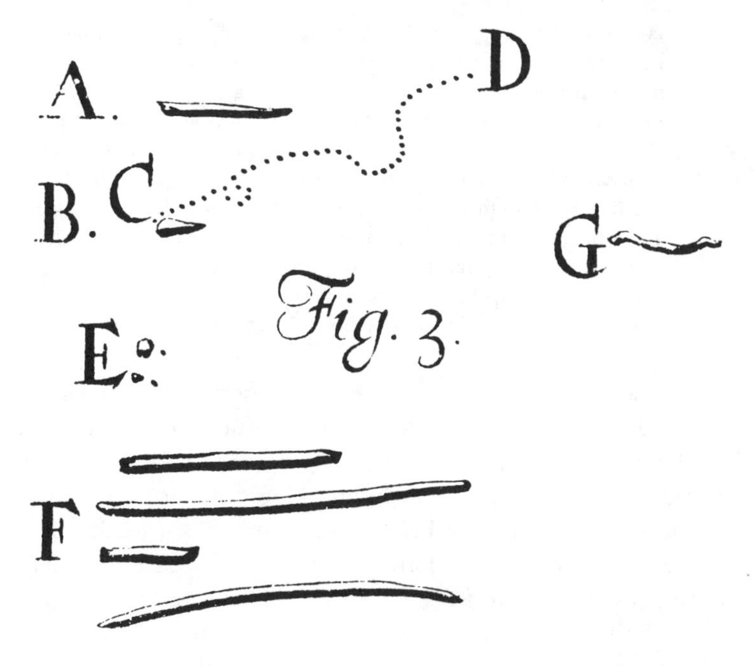

Figure 2 Leeuwenhoek's drawings of bacteria in the human mouth, published in 1684. Even from these crude drawings we can recognize several kinds of common bacteria. Those lettered A, C, F, and G are rod shaped; E, spherical or coccus shaped.

nought but what I had found in my own and other people's. I also took some of the matter that was lodged between and against his teeth, and mixing it with his own spit, and also with fair water (in which there were no animalcules), I found an unbelievably great company of living animalcules, a-swimming more nimbly than any I had ever seen up to this time. The biggest sort (whereof there were a great plenty) bent their body into curves in going forwards, as in Fig. G [see Figure 2]. Moreover, the other animalcules were in such enormous numbers, that all the water (notwithstanding only a very little of the matter taken from between the teeth was mingled with it) seemed to be alive. . . .I have also taken the spittle, and the white matter that was lodged upon and betwixt the teeth, from an old man who makes a practice of drinking brandy every morning, and wine and tobacco in the afternoon; wondering whether the animalcules, with such

> continual boozing, could e'en remain alive. I judged that this man, because his teeth were so uncommon foul, never washed his mouth. So I asked him, and got for answer: "Never in my life with water, but it gets a good swill with wine or brandy every day." Yet I couldn't find anything beyond the ordinary in his spittle. I also mixed his spit with the stuff that coated his front teeth, but could make out nothing in it save very few of the least sort of living animalcules hereinbefore described time and again. But in the stuff I had hauled out from between his front teeth (for the old chap hadn't a back tooth in his head), I make out many more little animalcules, comprising two of the littlest sort.

Leeuwenhoek's observations were all described (in the Dutch language) in about 300 letters, 190 of which were addressed to the Royal Society. His fame brought eminent visitors to his home in Delft (Holland), including kings and princes. There was a recent reminder of his extraordinary skills, revealed by a systematic search of his letters to the Royal Society.* Small envelopes attached to a few letters contained some of the specimens he had made and studied, including a thin slice of cork. The latter, prepared by hand by Leeuwenhoek, was examined with a contemporary research microscope and found to be "acceptable for laboratory use today." The specimen showed the honeycomb structure from which the term *cell* was derived in 1665.

---

*Ford, B. J., 1981, Leeuwenhoek's specimens discovered after 307 years. *Nature*, 292:407.

2

The Microbial Kingdom Has Many Subjects

Leeuwenhoek's observations demonstrated that microbes normally occur in very large numbers in human mouths and in our surroundings. Moreover, judging from their shapes alone, it was also clear that there were many different kinds. In addition to the single-cell microbes that abound in nature, there are also various types of multicellular forms; these have more complex internal structures and reproduce in more complicated ways than the single-cell microbes. Although a professional microbiologist can spend an entire career studying only one type of multicell microbe and its close relatives, there is no doubt that we can learn the basic essentials of microbes by primarily considering the single-cell types. Even the single-cell forms show different degrees of complexity, and this is the basis of separating them into two major groups, which has been referred to as "the great divide."*

**See* Morowitz, H. J., 1979, *The Wine of Life and Other Essays on Societies, Energy & Living Things*, St. Martin's Press, New York.

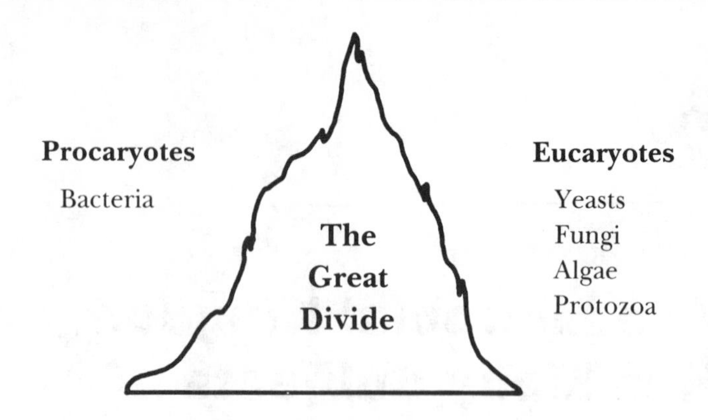

This separation is based mainly on whether or not the cell contains a well-defined nucleus of the kind seen in plant and animal cells (called *eucaryotes*). The nucleus of such cells is easily observable in the microscope as a distinct compartment of the cell that contains the genetic material in the form of filamentous structures called chromosomes. Yeasts, green algae,* and protozoa are types of single-cell eucaryotes that differ from one another in various ways, for example, in how they obtain energy for growth. On the left side of "the divide" are the procaryotes, that is, bacteria. These organisms have a comparatively simple anatomy, and do not have a distinct nucleus. The genetic material of bacteria (DNA) is of the same kind found in other living organisms, but it is organized in a different fashion, rather like a fuzzy blob floating in the cell interior.

WHY DIRECT OUR ATTENTION TO BACTERIA?

There are a number of good reasons for narrowing our main focus to the bacteria. First, they are the most numerous single-cell microbes on Earth. Soil is the largest repository of microbes; a typical number found in an upper layer of soil is on the order

*Note that some species of green algae are multicelled and macroscopic (that is, visible to the naked eye).

of 10,000,000 bacteria per gram (of dry soil). The population density varies significantly depending on the type of soil, season of the year, moisture content of the soil, and so on, but it is clear that bacteria are the predominant inhabitants. These bacteria comprise a great number of different species, the total number of which is still unknown. More species are constantly being discovered in soil and in other natural habitats. The microbiologist's "bible" is *Bergey's Manual of Systematic Bacteriology*. The eighth edition, published in 1974, is 3 inches thick and describes more than 1500 species of bacteria; the ninth edition has been expanded from one volume to four!

Another reason for emphasizing the bacteria is that as a group, they show the greatest diversity of "life styles." That is, they excel in their ability to grow and multiply in a wide range of environmental conditions. As they grow in ordinary and "extreme" environments, bacteria cause important chemical changes on the surface of the Earth. A recent text entitled *Geomicrobiology* by Prof. H. L. Ehrlich (Marcel Dekker, Inc., New York, 1981) begins: "The subject of geomicrobiology examines the role that microbes have played and are playing in a number of geological processes: for example, in the weathering of rocks, in soil and sediment formation and transformation, in the genesis and degradation of minerals, and in the genesis and degradation of fossil fuels." Most of the microbes discussed are bacteria.

If bacteria differed from other kinds of microbes in fundamental ways, it would clearly not be advisable to single them out for special emphasis. It is a fact, however, that the basic blueprint governing growth and development of bacteria is fundamentally the same as that of all other types of cells. Bacteria contain the same classes of essential biomolecules found in other cell types, and all cells produce their constituents by the same kinds of biosynthetic pathways. There are, of course, many differences among diverse cell types with respect to details of their growth mechanisms. The differences contribute to the special life history of each kind of cell, and exploring them has kept an army of scientists busy for many decades. However, advances made since the 1960s have dramatically demonstrated the great value of using bacteria as experimental systems for the study of major biological problems in all types of cells. Much of the contemporary molecular biology widely discussed in newspapers and news magazines is based on applying knowledge derived from the study of bacteria to the analysis of processes in plants and animals, including humans.

The general similarity of the basic ground rules of growth and reproduction in bacteria, other microbes, and all other types of cells is obviously not an accidental coincidence. This similarity is profound and is now widely agreed to reflect two important conclusions from decades of research in a number of scientific fields:

- Bacteria were the first forms of life on Earth.
- Bacteria evolved over the course of about 3.5 billion years, giving rise to the multitude of complex life forms that we see under the microscope or with the naked eye.

3

Some Microbes Prefer Life Without Air

INTRODUCTION

Following Leeuwenhoek's work in the mid-1600s, a century and a half elapsed before microbes were in the news again. In 1835, Agostino Bassi discovered that a microbe was responsible for an infectious disease of silkworms. Further research on the "germ theory of disease," however, was slow to develop notably because techniques for isolation of "pure strains" of microbes were not available. Moreover, the concept that microbes might be agents of chemical changes in their environments was not appreciated until the nature of fermentation processes was clarified. This was accomplished by Louis Pasteur who demonstrated that production of alcohol from sugar by certain microbes—"alcoholic fermentation"—was, as he put it, a "consequence of life without air." The idea that there were forms of life that did not require air (that is, oxygen gas) must have seemed strange to many in Pasteur's time. Indeed, it remains true to this day that the only organisms capable of growing and living indefinitely in the total absence of oxygen gas (called "anaerobes") all belong to the microbial kingdom.

FERMENTATION

It is an interesting and strange fact that knowledge of microbes and their activities remained at a standstill for more than a century after Leeuwenhoek's death. During this period, scientists were studying and debating various biological problems, including the process of *fermentation.* It would be more accurate to say *processes* of fermentation, because there are several kinds. The most familiar is the fermentation that gives rise to alcohol (and carbon dioxide), a phenomenon known to humans throughout recorded history and no doubt well before that. Another well-known fermentation process is the "spontaneous" souring of milk—the so-called lactic acid fermentation. In 1857, Louis Pasteur demonstrated that this seemingly spontaneous change was caused by bacteria that produce an organic* acid (lactic acid) from sugar during their growth in milk. The milk that is secreted into the udder of a healthy cow does not contain any microbes. When the milk is drawn, however, it invariably becomes contaminated with a variety of microbes, including bacteria and yeasts; these microbes are always present on the external surfaces of the udder and in dust particles floating in the air. To prevent (or delay) souring of milk, it can be heated briefly to a high temperature that kills most of the microbes present, a treatment known as *pasteurization.*

Lactic acid fermentation occurs not only in milk, but also in our muscles when we move or exercise. In both situations, the fermentation has the same function, that is, both lactic acid

*Organic acids and other organic compounds are chemical substances that contain atoms of carbon, hydrogen, and usually other elements such as oxygen.

bacteria and muscle cells derive the same kind of benefit from breaking down sugars to the smaller molecules of lactic acid. The nature of this benefit was eventually explained by analysis of the mechanism of alcoholic fermentation. After many decades of research, it became evident that fermentation occurs through a complicated series of chemical conversions that provide *energy* in a form that growing cells can use to fabricate and assemble their constituents (more details are given in Chapter 13). Fermentation was the first bioenergetic process to be understood from the standpoint of molecular chemistry, and the resulting clarification of the details of the fermentation process had a great impact on later developments in modern biology and medicine.

Sir Arthur Harden was awarded a Nobel Prize in 1929 for his research in unraveling the mechanism of alcoholic fermentation, and his 1914 book* on the subject eloquently summarized the historical background as follows:

> The problem of alcoholic fermentation, of the origin and nature of that mysterious and apparently spontaneous change which converted the insipid juice of the grape into stimulating wine, seems to have exerted a fascination over the minds of natural philosophers from the very earliest times. No date can be assigned to the first observation of the phenomena of the process. History finds man in the possession of alcoholic liquors, and in the earliest chemical writings we find fermentation, as a familiar natural process, invoked to explain and illustrate the changes with which the science of those early days was concerned. Throughout the period of alchemy fermentation plays an important part; it is, in fact, scarcely too much to say that the language of the alchemists and many of their ideas were founded on the phenomena of fermentation. The subtle change in properties permeating the whole mass of material, the frothing of the fermenting liquid, rendering evident the vigour of the action, seemed to them the very emblems of the mysterious process by which the long sought for philosopher's stone was to convert the baser metals into gold.

---

*Harden, A., 1914, *Alcoholic Fermentation*. Longmans, Green, New York.

Although ancient civilizations produced wine by fermentation and even used it as a staple in trading and payments of debts, it was not until the late seventeenth century that it was recognized that sweet-tasting materials such as fruits were particularly suited to undergo fermentation. Yet the explanation of the apparently spontaneous origin of fermentation and its propagation from one liquid to another remained shrouded in mystery. Further progress in understanding the essence of fermentation was slow, and a full century elapsed before establishment of the fact that during the process, sugar is converted to ethyl alcohol and gaseous carbon dioxide. The first major clue to the nature of the causative agent was not uncovered until 1837–1838 when, very remarkably, three independent observers concluded almost simultaneously that living yeast cells were responsible. At that time, yeast cells, which are invisible to the unaided eye, could be readily observed as spherical particles ("globules") using the microscope and were considered to be living organisms that probably belonged to the "vegetable kingdom." (Yeasts are normally found on the skins of fruits such as grapes, pears, and apples.)

The three investigators—Baron Charles Cagniard-Latour (a physicist), Theodor Schwann, and Friedrich Kützing—independently published their findings and interpretations at virtually the same time, and these were received with incredulity. In fact, for the next two decades the concept that fermentation is evoked by *living* microscopic organisms was ridiculed by leading chemists who were trying to explain the mechanism of the phenomenon. Adding insult to injury, an anonymous article appeared in the *Annalen der Pharmacie* (1839) under the title "The Mystery of the Alcoholic Fermentation Solved." The article stated that the problem of fermentation finally had been solved using a powerful microscope:

> Beer yeast broken up in water is resolved by this instrument [the microscope] into innumerable small spheres. . . .When placed in sugar water it can be seen that those are the eggs of animals;

> they swell, burst, and there develop small animals which multiply with incredible rapidity in a most unprecedented way. The form of these animals differs from that of the 600 species already described; it is the shape of a distilling flask. The tube of the stillhead is a kind of sucking snout covered internally with fine cracks; although teeth and eyes are not to be seen, one can distinguish a stomach, intestine, the anus (a rose-pink spot), and the organs of urine secretion. From the moment of emergence from the egg, the animals suck in sugar, which can clearly be seen in the stomach. It is immediately digested and the digestion is followed by excretion. In a word, these infusoria feed on sugar; they excrete from the intestine alcohol and from the urine organs, carbon dioxide. The urine bladder in the full condition is shaped like a champagne bottle. . . .

Historians have established that the authors of this wonderful farce were none other than the editors of the *Annalen der Pharmacie*: the famous chemists Justus Liebig and Friedrich Wöhler.

The chemists, of course, had their own ideas. One prominent theory held that fermentation was caused by a "body" called "the ferment" that somehow was formed as the result of air contacting plant juices that contained sugar. It was further supposed that the "ferment" had a remarkable property, namely, it was very unstable and could communicate its condition of instability to sugar which, as a consequence, fell apart to alcohol and carbon dioxide molecules. Even today, it is difficult to fathom this fanciful notion. Finally, in 1868 Pasteur settled the vexing question of the cause of the alcoholic fermentation to his own satisfaction, concluding that "alcoholic fermentation is an act correlated with the life and organization of yeast cells." In 1875 he made another great leap forward in connecting fermentation with the *energy* requirements of growing cells. He suggested that fermentation was the result of *life without gaseous oxygen* and that yeast as well as certain other kinds of microbes could obtain energy in the absence of oxygen gas by decomposing (fermenting) substances containing oxygen atoms in some combined form (not in the form of atmospheric oxygen gas). The air in the Earth's atmosphere contains about 20 percent oxygen gas, and one result of Pasteur's research was recognition of the important fact that there are many kinds of microbes that do not need this oxygen. He named such microbes *anaerobes*.

Well before Pasteur had switched the focus of his extraordinary experimental and conceptual abilities from chemistry to biology, alcoholic fermentation in the form of wine production and brewing of beer had become an established industry.* The founder of the Carlsberg Brewery in Copenhagen, J. C. Jacobsen, experienced a typical problem of brewers during the mid-nineteenth century: many brews tasted bad and had to be dumped into the sewer. Jacobsen was fascinated by the genius of Pasteur and lost no time in establishing (in 1875) the Chemical and Physiological Laboratory at his brewery, with this aim: "By independent investigation to test the doctrines already furnished by science and by continued studies to develop them into as fully scientific a basis as possible for the operations of malting, brewing and fermentation."

---

*A wall painting in an Egyptian tomb that had been sealed about 2000 B.C. shows that by that time the Egyptians had developed complicated methods of wine production. The painting shows grape vines trained to grow on trellises and being watered by hand.

4

Important Molecules in Microbes, Plants, and Animals

Why and how do yeasts and certain other microbes ferment sugar resulting in the production of alcohol? Interest in these and related questions led to a focus on chemical processes of microbes and other kinds of cells. This area of study in the late nineteenth century was called either physiological chemistry or biological chemistry and is now referred to as biochemistry. Solving the mechanism and function of the biological breakdown of sugar in the absence of air required many decades of research; biochemists were still at it in the 1930s. In many ways, the history of this great effort can be said to be the history of how the framework of modern biochemistry was erected. Almost every step forward in analyzing the problem required development of new techniques and led to new insights into how various kinds of cells obtain the raw materials needed to construct cell materials during growth and the energy required to assemble the "building blocks." It gradually became clear that the astonishing variety of "life styles" observed in the microbial universe reflects the different capacities of its inhabitants to use nutrients and obtain energy.

By now the reader must suspect that an understanding of microbial life requires at least an elementary appreciation of cell chemistry. In this chapter we will describe the most important chemical substances found in microbes. These substances are, in fact, the same classes of substances found in *all* types of cells.

The terms molecules, carbohydrates, fats, proteins, DNA, etc., are encountered daily in our lives: in newspapers, television advertisements, cereal box labels, and so on. If we wish to understand more of what is behind the headlines that deal with beneficial and harmful microbes, we must first examine some basic definitions and concepts of chemistry. Indeed without an elementary appreciation of the simplest aspects of chemistry, it is not really possible to comprehend the general features of fermentation, microbial growth, and ecology, or to understand how different kinds of microbes influence our agriculture, health, and comfort.

Since all matter is composed of chemical elements, it is understandable that operation of the machinery of living cells involves chemical processes. There are 92 naturally occurring elements, and about a dozen more have been made artificially. Those of particular importance in biology are conveniently grouped into three categories based on the relative amounts present in typical cells of microbes and other organisms.

Category I Elements that account for the major part of living matter. A useful mnemonic device for remembering these elements is "CHNOPS," pronounced as "schnapps," the name of a strong Dutch gin. The chemical symbols stand for carbon, hydrogen, nitrogen, oxygen, phosphorus, and sulfur, respectively.

Category II Four elements that occur in smaller, but significant, quantities, namely sodium, potassium, calcium, and magnesium.

Table 1 Elements Important in Microbial and Other Cells

Element	Symbol	Mass Units	Percentage of Human Body[a]
Category I			
Carbon	C	12	18.5
Hydrogen	H	1	9.5
Nitrogen	N	14	3.3
Oxygen	O	16	65.0
Phosphorus	P	31	1.0
Sulfur	S	32	0.3
Category II			
Sodium	Na	23	0.2
Potassium	K	39	0.4
Calcium	Ca	40	1.5
Magnesium	Mg	24	0.1
Category III[b]			
Iron	Fe	56	Trace
Zinc	Zn	65	Trace
Copper	Cu	64	Trace
Cobalt	Co	59	Trace
Selenium	Se	79	Trace
Molybdenum	Mo	96	Trace

[a]Approximate percentage of wet weight.
[b]Not all listed.

Category III The so-called trace elements. These are present in cells in very small quantities, but they have essential roles in cell chemistry.

Table 1 lists the elements of special importance in living matter, their relative weights (mass units), and their approximate abundance in the human body. The most prominent features of cell biochemistry are based on Category I elements. These usually occur in the form of chemical compounds. A *compound* is defined as a substance that consists of two or more elements united in definite proportions. In contrast, the term *molecule* refers to combinations of atoms in which the smallest unit that can exist still retains the properties of the original substance. Examples: H_2 is the formula for hydrogen gas, made up of molecules, each of which consists of 2 linked hydrogen atoms. CO_2 represents the gas carbon dioxide, a compound whose molecules consist of 3 linked atoms—one of carbon and 2 of oxygen. The size range of molecules encountered in cells is very large. For example, a molecule of water, written as H_2O, has three atoms (two of hydrogen and one of oxygen) and has a relative weight of 18 mass units. Glucose, written as $C_6H_{12}O_6$, has 24 atoms and weighs 180 mass units. Proteins, however, have a very large number of atoms; typical proteins weigh about 60,000 mass units, but some are up in the millions.

Modern concepts of the chemical properties of the elements began with important investigations by a remarkable English scientist, John Dalton (1766–1844). Dalton, a largely self-educated genius, developed the first quantitative atomic theory. He was able to demonstrate that atoms of different elements have different weights, and in his classic book *A New System of Chemical Philosophy* (Part II, published in 1810) he adopted a weight system based on the hydrogen atom (the simplest kind of atom). Thus: "The weight of an atom of hydrogen is denoted by 1, and is taken for a standard of comparison for the other elementary atoms." This convention sufficed for a long time, and for most purposes is still acceptable. For complicated reasons, however, the standard of reference now is the weight of a carbon atom. An atomic mass unit, now called a *dalton*, is one-twelfth the mass of a carbon atom.*

CHEMICAL BONDS

Dalton showed that when atoms of different elements combine to form compounds they do so in simple numerical proportions. To illustrate this he created new symbols for elements of dif-

---

*To be more precise we would have to say that a dalton is one-twelfth the mass of the major kind of carbon that occurs naturally: 98.89 percent of natural carbon consists of atoms that weigh 12 daltons; 1.11 percent of the atoms weigh 13 daltons.

ferent kinds, and his drawings indicated atoms linked together to form molecules. But what kind of "glue" holds the atoms together? Dalton had no idea of the forces involved. These forces of attraction or binding are now called *chemical bonds.* To simplify matters, you can imagine that each kind of atom has a certain number of arms or "hooks" each of which can link with an arm, or hook, of another atom. For example, if an oxygen atom has two hooks and a hydrogen atom one hook, we can visualize the combination of oxygen and hydrogen atoms to form water as follows:

```
           H—⊃
⊂—O—⊃  +          →      H ⁄ —O— ⁄ H
           ⊂—H
```

In diagrams, chemical bonds are usually depicted as a line (or a double line for a "double bond") that connects the symbols of two atoms. In this way the structure of water can be shown as H—O—H. Glucose (grape sugar) contains 24 atoms and is represented as follows:

```
      H     O
       \   //
         C
         |
     H—C—O—H
         |
 H—O—C—H
         |
     H—C—O—H
         |
     H—C—O—H
         |
     H—C—O—H
         |
         H
```

Such diagrams reflect the important principle that Category I elements can form a limited and characteristic number of chemical bonds: one bond for hydrogen, two for oxygen, and four for carbon. Compounds of carbon are of central importance in biology, and we must immediately distinguish between two classes of carbon compounds. On the one hand, carbon monoxide (CO) and carbon dioxide (CO_2) are designated *inorganic.* In contrast, compounds that contain chemical bonds between carbon and hydrogen atoms (and also other kinds of bonds) are called *organic.* Since each carbon atom can form only four bonds (and hydrogen can form only one), the simplest organic compound is methane:

```
     H
     |
  H—C—H
     |
     H
```

Carbon is extraordinary in that it is particularly versatile in combining (forming bonds) with other kinds of atoms. Accordingly, there is a very large range of sizes and varieties of molecules that contain carbon. It has been estimated that there are more than 500,000 possible combinations involving carbon, and in fact, there are far more different carbon-containing compounds known than the total number of compounds formed by all the other elements.

WE ARE WHAT WE EAT

An English physician, William Prout, was the first person to recognize, in 1827, that the principal foods used by humans "and the more perfect animals" were sugar, fats, and proteins; in his words these were "the saccharine, the oily, and the albuminous."* Nowadays, we refer to the sugary substances as

---

*The writings of Prout, who discovered the existence of hydrochloric acid in the stomach, are discussed by M. Teich in an interesting book that was the

carbohydrates. Thus, carbohydrates, fats, and proteins account for the major part of our foodstuffs, which we obtain mainly from plants and other animals. Other kinds of substances, such as DNA (and a related type of biomolecule called RNA) are present in plant and animal tissues, in relatively small amounts. Aside from water, our diets consist largely of

- Major components: carbohydrates, fats, and proteins.
- Minor components: nucleic acids (DNA and RNA) and mineral salts.

In addition to carbohydrates, fats, and proteins, small amounts of other kinds of nutrients are needed by humans and certain other organisms, notably vitamins. With a properly balanced diet, all of these materials are provided to us by natural foods, mainly from plant and animal sources. They are also all present in our own cells and in microbial cells. In other words, *all living cells contain the same kinds of chemical substances*; the relative proportions, however, may vary greatly.

With this background, we can now examine more closely several kinds of major cell constituents: carbohydrates, proteins, and fats. (The structure of DNA is discussed in Chapter 21.)

CARBOHYDRATES

This class of compounds contains C, H, and O atoms. Sugars are typical carbohydrates, and glucose ($C_6H_{12}O_6$) has already been given as an example. Note that in glucose, the ratio of H to O atoms equals 2, the same ratio as in water. This accounts for the origin of the term *carbohydrate*: carbon + "hydra" (water).

---

outcome of a course of lectures by eminent University of Cambridge biochemists: *The Chemistry of Life* (Lectures on the History of Biochemistry), J. Needham, Ed., Cambridge University Press, 1970; Teich's article, pp. 171–191, is entitled "The historical foundations of modern biochemistry."

Carbohydrates occur in many forms—as simple sugars (glucose, sucrose, etc.) and as more complicated structures in which glucose units are connected by chemical bonds. A molecule that contains many simple sugar units is also called a *polysaccharide*.

The most abundant polysaccharide on Earth is cellulose, found in the walls of plant cells. In cellulose, the glucose units are joined end to end, as in a linked chain. The chains are quite long and consequently form fibers, sometimes called *fibrils*. As plant cells grow, the fibrils are deposited in the cell walls, buried within a matrix of other materials. This arrangement strengthens the walls in the same way that concrete is reinforced by embedded metal rods. In wood, the cellulose fibrils are deposited in a material called lignin, and the manufacture of paper from wood consists essentially of separating out the cellulose and then matting the fibers together.

Glucose units can be joined together in other ways to produce polysaccharides with highly branched structures. There are two important examples of such large molecules (often referred to as macromolecules): *glycogen*, which occurs in animal muscle cells and in some microbes, and *starch*, which accumulates in certain plants (maize, potatoes, oats, etc.). If we compare the linear arrangement of glucose units in cellulose to a straight stretch of interstate highway, then we could compare the arrangement in starch and glycogen to the branching pattern in a mature tree (see Figure 3).

In brewing, the conversion of starch to fermentable sugar is achieved by "malting," which is a treatment using a preparation from sprouting barley. When barley sprouts in a warm and damp atmosphere, starch-degrading *enzymes* are formed in the germinating plant tissues. Enzymes are special proteins that accelerate chemical reactions. To obtain these enzymes in quantity, barley is permitted to begin sprouting and is then heated to a temperature that stops the germination but does not harm the enzymes. The resulting "malt" is then used to convert the starch to short chains of glucose units; most brewing yeasts cannot use anything larger than a chain of three glucose units.

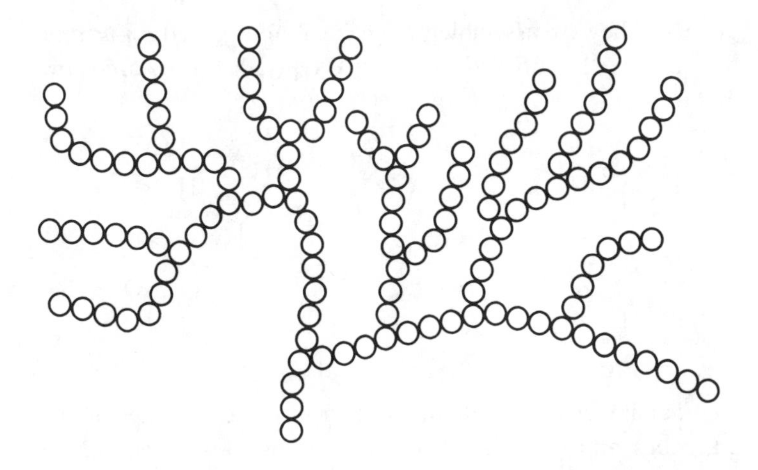

Figure 3 Schematic representation of part of a glycogen macromolecule. Each sphere represents a molecule of glucose. The glucose units are connected to each other by chemical bonds to produce a highly branched structure.

For commercial production of "grain alcohol" from corn, milled grain is mixed with water and then cooked to hydrate and gelatinize the starch. After treatment with barley malt, the mash is inoculated with an appropriate strain of yeast and the fermentation is complete in 40 to 60 hours. Finally, the alcohol in the fermented mash is distilled and purified.

Carbohydrates have two basic functions in living cells. Some serve as structural components or "building blocks," whereas others such as glycogen and starch, are energy reservoirs. To be used as energy sources, polysaccharides must first be broken down to individual glucose units, as in the malting process described above.

PROTEINS

Proteins are complex macromolecules that contain carbon, hydrogen, nitrogen and oxygen atoms (and sometimes other elements). They are assembled from small units called *amino acids*, which occur in about 20 forms; two typical varieties are shown here.

```
H\   /H                        H
   N                           \   /H
   |      //O                H   N
H—C—C                      |   |      //O
   |      \O—H          H—C—C—C
   H                        |   |      \O—H
                            H   H
```

All amino acids contain nitrogen atoms, and this explains why microbes and all other living organisms must have suitable sources of nitrogen atoms to grow. Proteins consist of long chains of amino acids hooked end to end by chemical bonds. In most proteins, the chains are folded in some particular way that depends on the particular sequence of the different amino acids in the chain (Figure 4). A typical protein contains about 500 amino acid units. If there are about 500 units in a chain, and a possibility of any of 20 *different kinds* of units at any one position in the chain, the number of possible sequences is astonishingly large (many, many millions).* A typical microbe contains approximately 3000 to 5000 different kinds of proteins. They all have different properties, which are determined by the particular sequence of amino acids in each protein.

Some proteins are designed to serve as structural cell components; these can be compared to structural beams used in house building. However, it is difficult to see how structural functions alone could explain why a cell would contain *thousands* of different kinds of proteins. In fact, the explanation of the large variety is that most cellular proteins are designed to participate in the numerous chemical reactions of metabolism, that

*Imagine that you could represent each type of amino acid by a letter of the English alphabet. Consider how many 500-letter "words" you could make with an alphabet of 20 letters.

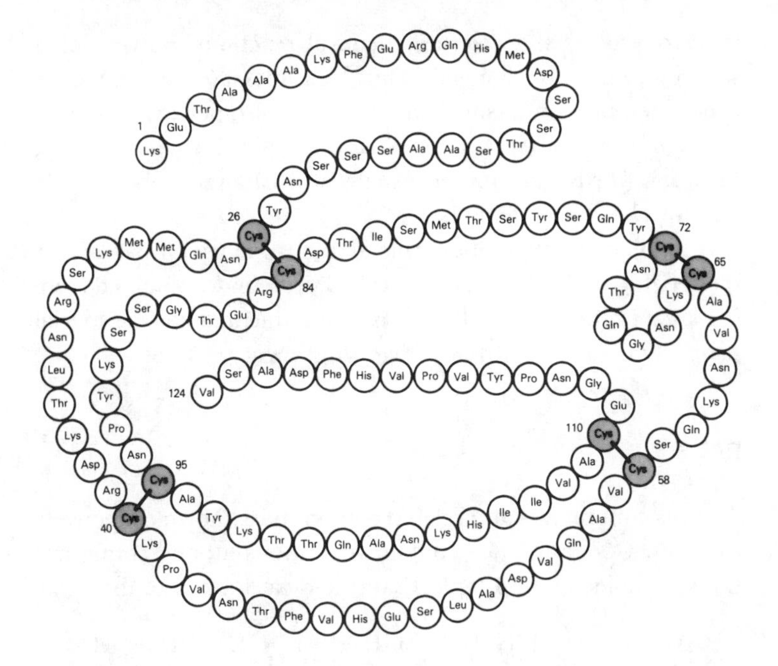

Figure 4 Structure of bovine ribonuclease, a relatively simple protein. Ribonuclease contains 124 amino acid "residues" connected to each other by chemical bonds in the sequence shown. The identity of each amino acid is indicated by a three-letter designation (Lys, lysine; Glu, glutamic acid; etc.). The shaded circles represent the amino acid cysteine, which is capable of making cross-connections, four of which are found in ribonuclease.

is, the biochemical processes by which cells obtain energy and produce their characteristic constituents. The machinery of metabolism is a gigantic complex of thousands of integrated chemical reactions that must proceed in finely tuned (regulated) fashion. Each kind of chemical reaction is catalyzed (accelerated) by a particular kind of protein. Such proteins are known as *enzymes*. A typical microbe contains several thousand kinds of enzyme catalysts. The term *catalyst* means an agent that speeds up the velocity of a chemical reaction without changing the nature of the reaction or adding any energy. Enzymes, like other kinds of catalysts, accelerate chemical reactions, but in doing so they remain unchanged. Thus, enzymes are very effective even when present in small amounts. It is also pertinent to note that enzyme catalysts are extremely specific in their activities; in other words, an enzyme usually will catalyze only a single kind of chemical reaction.

A familiar example of enzyme action is the use of papain for tenderizing meat. Papain is a so-called proteinase, obtained from the latex of the papaya plant. It is an enzyme that has the special property of breaking down (digesting) meat proteins.

FATS

All cells contain fats and related compounds, known collectively as *lipids*. Generally they contain repetitious atomic configurations; a molecule of a typical saturated fat looks like this:

```
  H   O   H H H H H H H H H H H H H H H
  |   \\  | | | | | | | | | | | | | | |
H-C-O-C-C-C-C-C-C-C-C-C-C-C-C-C-C-C-C-H
  |       | | | | | | | | | | | | | | |
  |       H H H H H H H H H H H H H H H
  |
  |   O   H H H H H H H H H H H H H H H
  |   \\  | | | | | | | | | | | | | | |
H-C-O-C-C-C-C-C-C-C-C-C-C-C-C-C-C-C-C-H
  |       | | | | | | | | | | | | | | |
  |       H H H H H H H H H H H H H H H
  |
  |   O   H H H H H H H H H H H H H H H
  |   \\  | | | | | | | | | | | | | | |
H-C-O-C-C-C-C-C-C-C-C-C-C-C-C-C-C-C-C-H
  |       | | | | | | | | | | | | | | |
  H       H H H H H H H H H H H H H H H
```

In some lipids, the "monotony" is relieved by the presence of an atom of phosphorus or nitrogen. Fats and other lipids differ from carbohydrates in that they contain a much smaller proportion of oxygen, and this is part of the reason why lipids are "oily" and do not dissolve in water. Lipids have two principal biological functions: in some kinds of cells, fats are stored for later use as sources of energy and small "building block" molecules, but more importantly, lipids are important constituents of membranes. All cells are bounded on their external surface by a lipid-containing membrane, frequently reinforced by other kinds of cell wall materials. In many microbes, the surface membrane extends into the body of the cell, and such membrane projections frequently have important roles in energy metabolism.

METABOLISM

In the process of digestion in an animal, large nutrient molecules are broken down into simpler, smaller units which are then reassembled to produce the characteristic components of that animal. Proteins and nucleic acids, in particular, are frequently "species specific"; that is, each animal species makes its own kinds. The production of these and other cell constituents, and all other chemical changes in cells, is referred to as *metabolism.* We all know from experience that metabolism also involves interconversions of dietary components, for example, the transformation of carbohydrates (sugars, etc.) into fats. This brings us to an important fundamental principle: the assembly (or reassembly) of small molecular units into cellular carbohydrates, fats, proteins, etc., *requires substantial inputs of energy.*

The growth of microbes (as opposed to animals) rarely involves a digestive phase; rather, microbes are typically dependent on supplies of certain small molecules in their environment. Assembly of these small molecules into microbial proteins, nucleic acids, etc., is the name of the microbial growth game, and as in the metabolism of all organisms, this requires energy. Microbes excel in the number of alternative ways they can obtain growth energy. The "life style" of a microbial species is frequently a reflection of the means by which the organism generates its energy needs. Some typical examples include fermentative anaerobes, exemplified by yeast and lactic acid bacteria; photosynthetic bacteria, which use light as their energy source (as green plants do); and *aerobic* microbes, organisms that must have oxygen gas (air) for their bioenergetic mechanisms. In addition, many microbes use unique variations of these bioenergetic themes.

5

How Microbes Are Isolated and Identified

Microbes catalyze numerous environmental chemical changes while they are pursuing their own purposes (namely, the fabrication of new microbial cells). These effects could not be intelligently interpreted until methods were developed for separation of natural mixtures of diverse microbial types into "pure" strains. Only then was it possible to determine by deliberate experiment, in the laboratory, the capabilities of individual kinds of microbes. With this knowledge (still being acquired), analysis of natural events in which myriads of microbes participate became feasible. The basic methodology of microbiology was firmly in place by the mid-1940s, but at the time no one could foresee the enormous dividends microbiological techniques and the study of bacteria and their viruses would bring to analysis of the fundamental processes of animal and plant cells during the 1970s and 1980s.

The false notion that living organisms arise from inanimate materials by "spontaneous generation" was entertained by "natural philosophers" for many centuries. After it had been shown decisively that this could not be true for higher organisms of any kind, the battleground of debate on this question shifted to microorganisms. Pasteur devised ingenious ways of proving that microbes also do not arise by "spontaneous generation," but that they are produced instead from other microbial cells. One of the important procedures used in Pasteur's studies was preliminary destruction of all living microbial cells in nutrient fluids by heating, typically by boiling solutions for 15 to 20 minutes. He demonstrated that when nutrient fluid in a flask is treated this way, and the neck then drawn out and bent as shown below, the fluid remains free of microbial growth indefinitely. In this arrangement, the gases of air communicate freely with the fluid in the flask, but dust particles cannot ascend the bent tube; consequently, microbe-laden dust particles cannot make contact with the fluid.

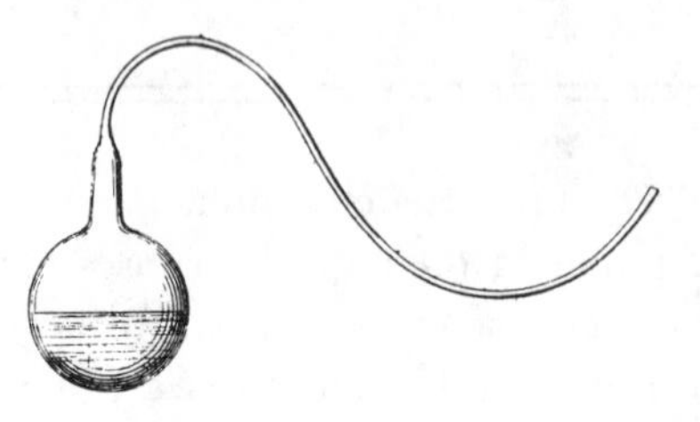

If, however, the neck of the flask is cut off, so that dust particles can drop into the fluid, the latter will soon be teeming with microbial growth. This clever, yet simple, experiment disposed of vague arguments to the effect that heating could destroy an air component necessary for the "spontaneous generation" of microbes.

Heating effectively kills virtually all kinds of microbes, but on occasion it is found that some can survive this harsh treatment. Usually, the microbes that survive exposure to elevated temperatures are species of bacteria that are able to produce specialized structures called endospores (see Chapter 7). The latter are unusual life forms in that they are extraordinarily resistant to adverse environmental conditions. It appears that endospores can remain alive but dormant for long periods of time. Under suitable circumstances, endospores can germinate, giving rise to "ordinary" bacterial cells. Some species of disease-producing bacteria form endospores, and their long-term persistence in soils or other natural reservoirs can pose public health problems.

Pioneering research during the mid-nineteenth century by Louis Pasteur, Robert Koch, and others made it clear that humans are constantly exposed to microbes of many kinds. In 1860 it no doubt seemed hard to believe that large numbers of microbes float around on invisible dust particles in the air. Yet this was the basis for one of Pasteur's most dramatic experiments. He left on a journey from Paris with 73 sealed flasks each containing a clear fluid that was rich in the nutrients needed for growth of microbes, called the *growth medium*. Before starting, the flasks were sterilized, that is, heated to a temperature high enough to kill all microbes that happened to be in the materials used to prepare the growth medium. The flasks, carried by a mule, were to be opened briefly, one at a time, at different stops along Pasteur's journey. If microbe-laden dust particles dropped into a flask, the nutrient fluid would look cloudy the next day due to growth and multiplication of the microbes. Pasteur reasoned that the higher he went, the purer the air and the fewer the microbes. The route led eventually to the Chamonix glacier in the French Alps, and the results were as he expected. It is known that he would have liked to charter a balloon to prove that he could finally get to a place in the atmosphere where there would be *no* microbes at all.

Other experiments of Pasteur and his contemporaries clearly indicated that in our surroundings there is a considerable variety of microbes having many different properties. A goal of this early research was to discover a method of determining the unique properties and capabilities of the different kinds. Methods had to be devised to separate the mixtures of microbes that occur naturally into individual pure strains, that is, into populations that contain cells of only a single kind of microbe. Such populations are now called "pure cultures." Most of Pasteur's research was not done with pure cultures but rather with cultures that were highly enriched in a particular kind of microbe.

ENRICHMENT CULTURES

It sometimes happens that one particular kind of microbe grows abundantly in a certain natural ecological niche by outcompeting other microbes for available nutrients. The unsuccessful microbes die off gradually, whereas cells of the successful strain become the predominant type in the population; this is known as a "bloom." Isolation of the predominant cell type from a bloom is much easier than from a random mixture of many microbial species. Blooms occur because the chemical and physical conditions favor more rapid growth of a particular microbe; in effect, this microbe enjoys a *selective* advantage over other types in certain special circumstances that develop in natural environments. However, we need not wait and search for such natural events because these selective conditions can be deliberately arranged in the laboratory. Setting up enrichment (selective) cultures in the laboratory is, in fact, a much-used approach for facilitating isolation of different species of microbes. A concrete example may be instructive.

Certain soil bacteria have the special ability to obtain gaseous nitrogen (N_2) from the atmosphere and use it as their source of nitrogen atoms for growth. (See Chapter 10 for more discussion of nitrogen.) Most bacterial species are unable to do this. The "fixation" of N_2 by soil bacteria is important for crop plant productivity; N_2-fixing bacteria enrich soil with forms of nitrogen utilizable by plants. Thus, there is great interest in the isolation and study of such organisms. Detection of the N_2-fixing bacteria in a soil sample is done by adding soil to a "nutrient soup" (the growth medium) which contains all the chemicals needed for growth except for a source of nitrogen. Air (which contains 80 percent N_2) or pure N_2 gas is bubbled through the medium. The N_2 fixers obviously have a selective advantage under these conditions and become highly enriched in the population. They can then be isolated more readily. The same principle can be used to enrich for other microbial species with different special capabilities.

Enrichment cultures rarely become pure cultures; there are always a few other types that eke out an existence living on nutrient "scraps" discarded by the main population. Thus, a method is needed for ensuring that all the cells in a culture are of the same kind (a "clone"). The method most widely employed can be used either with enrichment cultures or with natural mixtures of microbes. It is based on the principle that a pure culture consists of cells that are all derived from a *single* ancestral cell. If a single cell can give rise to a large population of new identical cells, the reproduction process is obviously *asexual*. In other words, genetic input from two kinds of parental cells, as occurs in higher forms of life, is not essential for growth of microbes. Indeed, they ordinarily multiply in this fashion, sometimes called "vegetative" growth. Under suitable conditions, bacteria and other microbes also can reproduce by mechanisms that involve gene exchange between individual cells; several kinds of such "sexual recombination" in bacteria (and other microbes) have been discovered and are considered in Chapter 21.

A NEW METHOD—PURE CULTURES

In 1881, developments of major significance for the future of microbiology were made by Robert Koch (1843–1910), a German physician who served as a surgeon in the Franco-Prussian war. Koch received a Nobel Prize in 1905 in recognition of his research contributions to medical microbiology; he discovered and isolated the bacterial species that cause tuberculosis and cholera. These and other discoveries depended on a new method that Koch devised for obtaining pure cultures.

Koch prepared a solid growth medium by incorporating 2.5 to 5 percent gelatin into the nutrient soup. The gelatin medium was sterilized by heating, and while still in the liquid state, was poured onto a thin glass slide (about 1 × 3 inches in size). After the gelatin had set (become solid), the solid, transparent medium was inoculated with a mixture of bacteria as follows. A fine metal wire was sterilized in a flame and after cooling was dipped into the source of bacteria. The wire tip was then drawn rapidly and lightly over the surface of the gel, in a pattern like that shown here. The numbers refer to the sequence of streaks.

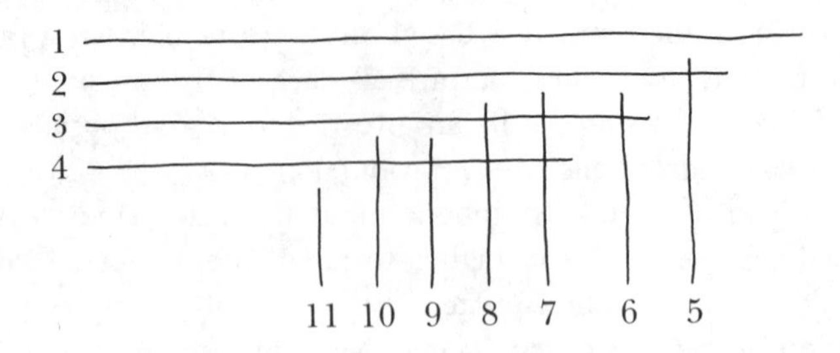

The slide was then placed in a warm incubator. When observed one to two days later, dense, opaque masses of cells were seen along the initial streaks. Invariably, however, along some of the last streak lines (e.g., streaks 5–11), round *colonies* of cells had developed, well separated from one another. Colonies of bacteria are usually about 1 to 2 millimeters in diameter. How is this result explained?

When the wire tip is dragged along streak no. 1, large numbers of bacteria are sloughed off. This is also true for streak no. 2 (and sometimes streak no. 3), but eventually the bacteria are rubbed off the needle *one by one*. During incubation of the slide, each cell grows and divides into two cells, the latter grow and divide, yielding 4 cells, etc. In a relatively short time a surprisingly large number of cells accumulate (Table 2). This results in a continuous mass of cells along the initial streak (and along part of the second streak), but at any point where a *single*

Table 2 Number of Cells Produced in 24 Hours from Multiplication of a Single Cell (Doubling Time: 1 Hour)

Time (Hours)	Number of Cells
0	1
1	2
2	4
3	8
4	16
5	32
6	64
7	128
8	256
9	512
10	1,024
11	2,048
12	4,096
13	8,192
14	16,384
15	32,768
16	65,536
17	131,072
18	262,144
19	524,288
20	1,048,576
21	2,097,152
22	4,194,304
23	8,388,608
24	16,777,216

cell has come to rest, all of its direct descendents remain localized in the form of a single round colony. The colony represents a *clone*—a pure culture. In some instances, to be sure that a colony was derived from only a single cell, the procedure is repeated again using cells from a single colony on the first slide as the inoculum.

Before this simple procedure was devised, the methods in use for obtaining pure cultures were tedious and unreliable. Koch summarized his great innovation as follows:

> The peculiarity of my method is that it supplies a firm and, where possible, transparent pabulum; that its composition can be varied to any extent and suited to the organism under observation; that all precautions against the possibility of after contamination are rendered superfluous; that subsequent cultivation can be carried out by a larger number of single cultures of which of course only those cultures which remain pure are employed for further cultivation; and that, finally, a constant control over the state of the culture can be obtained by the use of the microscope.

The new technique was a sensation and opened the door to great advances in biology and medicine. The original glass slide technique was soon improved by R. J. Petri, one of Koch's assistants. He poured the liquid nutrient gelatin medium into a sterile round glass dish (about 3 inches in diameter with a rim 0.5 inch high) that had a glass cover (Figure 5). This arrangement, essentially like a glass pillbox, had the advantage that one could examine the surface of the medium to see how growth was developing without exposing the culture to airborne dust particles laden with diverse microbes. These dishes, now called "Petri dishes" after their inventor, are used in enormous quantities in thousands of research and hospital laboratories around the world.

Koch and his colleagues did encounter two problems with their solid gelatin media. Certain bacteria decompose gelatin (which is a protein obtained from animal tendons) and thereby can affect the consistency of the solid medium. A more significant problem for Koch, who was mainly interested in bacteria that cause human diseases, was that gelatin does not remain solid at body temperature (98.6°F), and this is the optimal temperature for growth of many pathogenic bacteria (those that cause disease).

The problem was solved by the wife of one of Koch's coworkers, Walther Hesse. Fanny Hesse, the daughter of a German immigrant to the United States, suggested to her husband that agar be used in place of gelatin. Agar is a complex polysaccharide obtained from algae and has long been used for

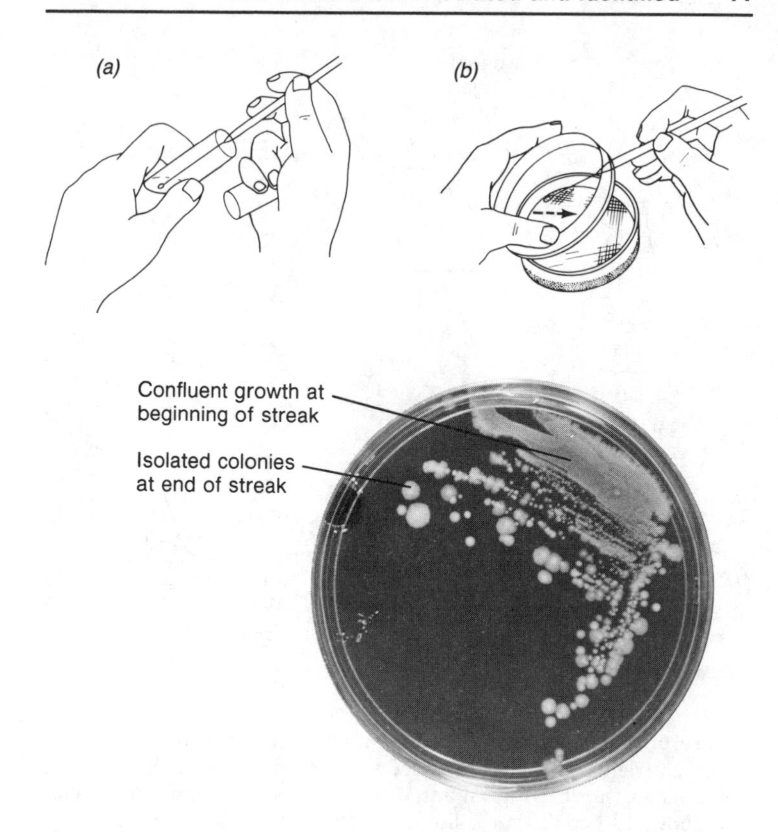

Figure 5 Procedure for isolating clones of microbes by streaking a droplet of cell suspension on a solid growth medium. (a) Loopful of inoculum is removed from tube. (b) Streak is made over a sterile agar plate, spreading out the organisms. (c) Appearance of the streaked plate after incubation. Note the presence of isolated colonies. It is from such well-isolated colonies that pure cultures usually can be obtained.

cooking purposes such as preparing fruit and vegetable jellies and thickening soups. Mrs. Hesse's recipes had come to her mother from Dutch friends, former residents of Java where such use of agar was common. Koch rapidly adopted agar as a solidifying agent (Figure 6). It was much superior to gelatin because at a temperature of about 42°C (108°F), molten agar

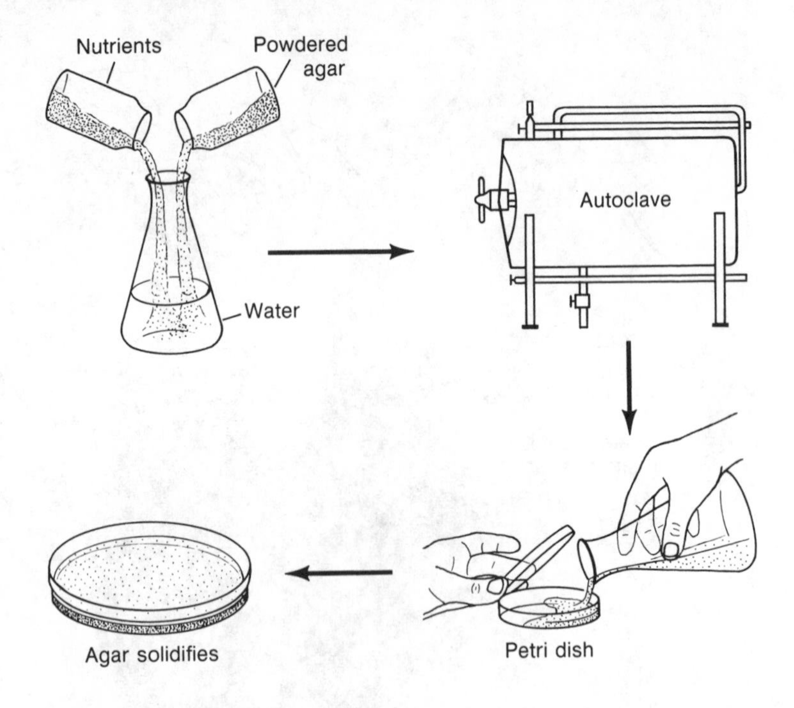

Figure 6 Preparation of Petri dishes containing agar culture medium. Nutrients and agar are added to water in a large flask. The flask contents are sterilized in an autoclave, then poured into Petri dishes and allowed to cool; agar solidifies at 42°C (108°F).

sets into a stiff, relatively transparent gel that does not melt at body temperature.

THE USES OF PURE CULTURES

With the availability of a simple method for obtaining pure cultures of bacteria and other microbes, the library of microbes in captivity expanded rapidly and has continued to do so. It was necessary to give them names and to describe them (for example, their shapes and sizes under the microscope and what their colonies on agar looked like). As for all other living or-

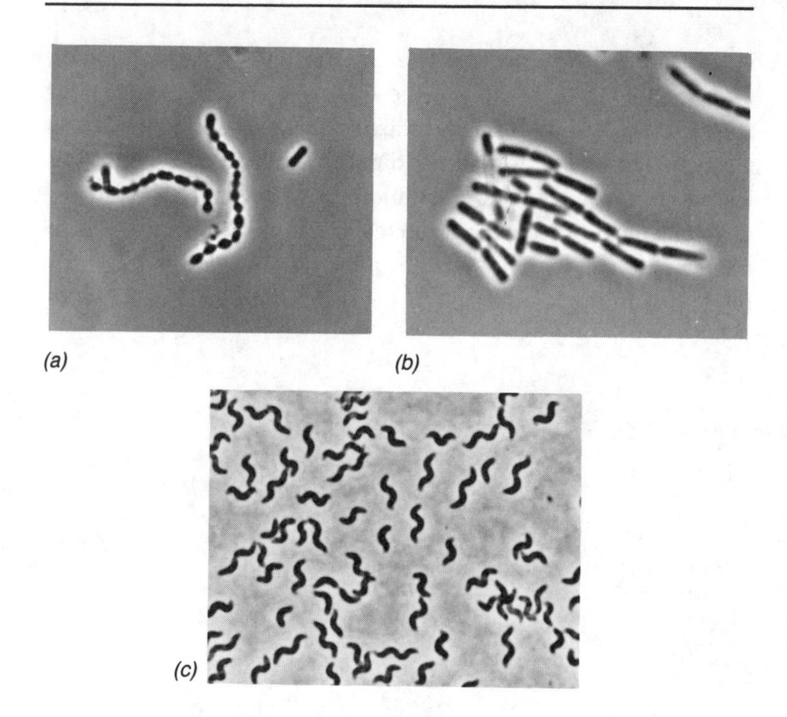

Figure 7 Bacteria of different morphology. (a) Coccus shaped (spherical); *Streptococcus*. (b) Rod shaped; *Bacillus*. (c) Spiral shaped; *Rhodospirillum*.

ganisms, each microbe was given a genus and species name. For example, the bacterium that causes tuberculosis was named *Mycobacterium tuberculosis*, and a common lactic acid-producing bacterium was designated *Streptococcus lactis*. Some were named after the scientist who first isolated the organism, others according to the cell shape, and still others after some outstanding property of the organism (for example, *Methanobacterium* for certain bacteria that produce methane). Some examples of different morphological types of bacteria (those having different shapes) are depicted in Figure 7.

Once a pure culture is available, it is possible to grow the microbe in quantity so that detailed studies of its various properties can be made. From such knowledge, we can assess its significance with respect to its interactions with plants, animals, and other microbes; its possible roles in chemical conversions that occur constantly on the Earth's surface; and its potential for use in biotechnology. In addition, recent research indicates that a comprehensive understanding of the structures of macromolecules (proteins and nucleic acids) of different kinds of bacteria may eventually enable us to piece together the early history of life on Earth.

6

The Care and Feeding of Microbes

At the time the agar-streaking procedure was introduced (and for the following 30 years), little was known about the specific nutrient requirements of different types of microbes. Thus, older media recipes usually included complex supplements that were assumed (justifiably) to be very rich in nutrients of various kinds; for example, extracts of soybean meal or of yeast cells. Many microbes can, indeed, grow in a solution made by simply dissolving 0.5 gram (about 0.02 ounces) of dry yeast extract powder in 100 milliliters (about 0.21 pint) of water (the yeast preparation is made by drying the water-extractable portion of a paste of yeast cells). The exact composition of such complex media is not known. With the passage of time, it became evident that in many instances the chemical activities of bacteria and other microbes can be significantly affected by the presence or absence of particular nutrients in the growth medium. This led to the development of *synthetic* media, whose compositions were completely specified by using only pure chemicals in preparing the recipes.

Since there is an extraordinary degree of metabolic versatility in the microbial world, thousands of culture media recipes have been proposed and used. For purposes of illustration, we will consider one important bacterium that is widely used in microbial research and happens to have relatively simple nutrient requirements, namely, *Escherichia coli*.

ESCHERICHIA COLI AND ITS GROWTH REQUIREMENTS

Theodor Escherich (1857–1911) was a pioneering Viennese physician considered to be one of the leading pediatricians of his day. Aside from his busy medical practice, he was active in fundamental research that focused on bacterial flora in the intestine of infants and the roles of bacteria in the physiology of digestion. Escherich demonstrated that certain strains of what he named *Bacterium coli* could cause infant diarrhea and also gastroenteritis (inflammation of the membranes of the stomach and intestine). By *strains*, we mean pure cultures of the same organism (see Chapter 5) isolated at different times or from different individuals. Strains frequently differ from each other in minor ways, but they are clearly recognizable as belonging to the same biological grouping.

In honor of Escherich, the genus *Bacterium* was later named *Escherichia* (commonly abbreviated to *E.* when used with a species name also). *E. coli* is found in the large intestine soon after birth and constitutes part of the normal bacterial flora throughout life. (The total number of bacteria excreted each day by an adult is normally between 100 billion and 100 trillion!) At any one time, the feces of an adult contain a number of different strains (about two to ten or more) of *E. coli*. Each of us becomes adapted to our own strains, and it frequently happens that ingestion of "foreign" strains causes minor disturbances of the gastrointestinal tract (such as in traveller's diarrhea). The extensive study of *E. coli* has had a profound influence on the spectacular progress made in biology and some aspects of med-

Table 3 Composition of a Liquid Medium for Growing *Escherichia coli*

Component	Chemical Formula	Category I Elements	Grams per Liter of Medium
Glucose	$C_6H_{12}O_6$	C, H, O	5
Ammonium chloride	NH_4Cl	H, N	2
Sodium phosphate	Na_2HPO_4	H, O, P	6
Potassium phosphate	KH_2PO_4	H, O, P	3
Sodium chloride	NaCl	—	3
Magnesium chloride	$MgCl_2$	—	0.01
Sodium sulfate	Na_2SO_4	S, O	0.026
Water	H_2O	H, O	(1 liter)

icine during the past 40 years. There is little question that we know more about *E. coli* than about any other living organism, and it is now one of the principal microbes used for genetic engineering in biotechnology.

Escherichia coli grows rapidly in synthetic media of simple composition. The recipe given in Table 3 has been widely used for growing batches of *E. coli* cells for research purposes. Note that some of the elements discussed in Chapter 4 are missing from the recipe, namely, calcium (Category II) and trace elements (Category III). These elements are required by *E. coli*, but sufficient quantities are present in any ordinary water supply to satisfy the nutritional needs.

Before inoculating the sterilized medium with a starter culture of *E. coli*, the investigator must make one more decision. The medium can either be purged of atmospheric oxygen with a (sterile) gas such as helium or argon to keep the culture anaerobic, or alternatively, it can be continuously bubbled with sterile air during incubation of the culture. Unless there is some special reason for growing the cells in the absence of oxygen gas (O_2; O=O), the second course is likely to be followed. *Escherichia coli* can obtain energy for growth by fermenting glucose (in the absence of O_2), but it can also get its energy through the process of *aerobic respiration* of glucose. This mechanism requires gaseous oxygen and is much more efficient than fermentation.

The energy source in the medium is provided by the addition of glucose, which contains chemical energy in the form of the C—H bond and other chemical bonds. To be useful for growing cells, this bond energy must be transformed into a specialized kind of energy-rich molecule, called ATP, that can be put to work in the synthesis of cell constituents such as proteins and nucleic acids. ATP can be thought of as a common "currency" for the transfer of energy to the various cell systems that require energy for their activities. It is called a currency in the sense that energy can be stored in the ATP molecule and then used for performing a function, just as money is a currency that can be stored in a bank and then withdrawn for various uses. The nature of ATP currency and how it is constantly regenerated will be considered in Chapter 13. For now, it is sufficient to say that energy conversion from the energy of chemical bonds of glucose to ATP currency is far more efficient in aerobic respiration than in fermentation—38 "units" of currency can be made for each glucose molecule that is metabolized by aerobic respiration and only 2 units for each glucose molecule that is fermented.

Aerobic respiration is clearly an advanced form of energy conversion. Indeed, all animal life is dependent on this kind of biological energetics and could evolve only after the Earth's atmosphere contained sufficient gaseous oxygen. It is the general consensus that after the Earth was formed, its atmosphere was anaerobic for about two billion years. The first life forms must have been anaerobic, fermentative microbes. Eventually, photosynthetic bacteria that could produce O_2 from water (now known as cyanobacteria) made their appearance, and later, green plants. With O_2 accumulating in the atmosphere, the road was paved for aerobic organisms that could use O_2 for more efficient generation of cellular energy currency (ATP).

The minimal nutritional requirements of *E. coli* reflect that this bacterium has very well developed biosynthetic capacities.

In other words, it can produce all cell constituents from simple sources of carbon, nitrogen, and mineral salts. However, some microbes do not have the necessary enzyme catalysts for making one (or more) of the building block units (such as amino acids) from which cell constituents are assembled. In this case, if the amino acid building block is not added to the growth medium, the microbe will be unable to grow. Certain bacteria have numerous biosynthetic deficiencies and to grow such organisms many "growth factors" must be added when synthetic media are prepared. Before these growth factors were identified, microbiologists used complex supplements of the sort already noted (extracts of yeast, soybean meal, etc.). This practice is still frequently used for experiments in which it is not essential to grow the microbe in a completely synthetic medium.

STORAGE OF MICROBES

Koch's procedure for isolating pure cultures by streaking bacteria on "nutrient agar" was quickly adopted by microbiologists. Before long, important discoveries were made in identifying microbial species that were the agents of either beneficial or deleterious effects on animal and plant life. Each new species was studied to determine such characteristics as shape and size of cells, special nutrient requirements, and outstanding metabolic features. The descriptions were published in technical journals, and eventually the need for identification manuals (such as *Bergey's Manual of Systematic Bacteriology*) became apparent. Microbiologists in various countries were busy isolating pure cultures of microbes from soil, airborne dust particles, natural waters, and plant and animal surfaces. In each case it was necessary to determine whether or not they were the same as strains isolated elsewhere. Since different strains of the *same* species frequently show detectable minor differences in some properties, it also became evident that direct comparisons with standard reference strains were often necessary. Cells of the standard ("type") strain could be kept alive, but not growing, in the form of colonies or streaks on agar plates stored in the refrigerator (at 4°C, or 39°F).

Experience gradually showed that at 4°C, cells usually remained viable for many months and sometimes years, that is, alive and capable of growing rapidly when inoculated into a suitable nutrient medium and incubated at the appropriate temperature. Microbiologists also learned that they could lose valuable pure cultures if errors were made in preparing growth media or if refrigeration and/or incubator equipment went out of control for one reason or another. Obviously, reliable collections of reference strains were needed to facilitate ongoing research, and these were established in several countries. The American Type Culture Collection (ATCC) is the largest collection in the world, and contains the most diverse assortment of known microbes. It was founded in 1925 by a committee of microbiologists and is located in Rockville, Maryland. The ATCC is a private, nonprofit corporation governed by a Board of Trustees that consists of scientists who represent professional societies such as the American Society for Microbiology and the American Society of Tropical Medicine and Hygiene. Financial support comes from funds provided by government agencies, contributions from scientific societies, and fees received for cultures provided to scientists in various kinds of laboratories.

The ATCC maintains more than 34,000 strains representing about 10,000 different species of cells, mainly microbes. Currently, about 500 new species are added to the collection each year. (See Appendix II for some recent additions.) For a relatively small fee, a sample of any culture can be purchased by anyone who has a legitimate need (for teaching or research purposes).* The scientific staff of the ATCC is constantly en-

*An example of an attempt at illegal use of this resource is described in an article headlined "2 Charged in Deadly Bacteria Plot" that appeared in the *St. Louis Post-Dispatch* of November 25, 1984. In the fall of 1984, the ATCC re-

gaged in research aimed at improving cell preservation techniques and increasing knowledge of the properties of different species. In 1982, the ATCC distributed more than 47,000 cultures.

CELL PRESERVATION TECHNIQUES

It has been known for some time that if microbial cells are completely dehydrated under suitable conditions, the dried cells can remain alive for long periods. A common technique now used for preserving microbial cells is based on freeze-drying ("lyophilization"). Cells of a pure culture are added to a small volume of sterilized milk or some other sterile nutrient fluid in a glass ampoule. The cell suspension is frozen, and while in this state, water is removed by applying a vacuum until the sample is completely dry. The ampoule is then sealed by melting (fusing) the glass neck of the container. The ATCC has an inventory of more than 1,000,000 such ampoule cultures (or similar vials) that are stored at low temperatures (Table 4). Freeze-dried cultures are usually kept at −60°C (−76°F).

An alternative procedure involves storage of cells at much lower temperatures. In this method, the cells are suspended in diluted growth medium that is supplemented with stabilizing chemicals, and the container is immersed in liquid nitrogen (N_2) at a temperature of −196°C (−321°F). A special storage room at the ATCC contains numerous stainless steel tanks, each of which holds 40,000 culture vials suspended in liquid nitrogen.

The storage facilities at the ATCC are under 24-hour surveillance by electronic monitors; empty spare deep-freezes are

---

ceived requests for cultures of the bacteria that cause tetanus and botulism, under circumstances that aroused suspicion. The Federal Bureau of Investigation was contacted, and a package containing harmless materials substituting for the requested bacteria was sent. FBI agents arrested two suspects when they arrived to pick up the shipment. According to the news report, "Authorities say they don't know why two men allegedly planned to smuggle enough tetanus and botulism bacteria into Canada to 'wipe out a whole city'."

Table 4 Examples of Bacteria in the ATCC and Similar Collections

Bacterium	Outstanding Characteristics
Azotobacter vinelandii	Fixes atmospheric N_2
Bacillus anthracis	Causes anthrax in animals and humans
Bacillus polymyxa	Produces the antibiotic polymyxin
Clostridium acetobutylicum	Produces useful solvents (acetone and butanol) from sugars
Clostridium botulinum	Produces the toxin responsible for botulism
Clostridium tetani	Produces a paralytic toxin that causes tetanus
Erwinia carotovora	Causes "soft rot" of vegetables
Heliobacterium chlorum	A photosynthetic bacterium that fixes atmospheric N_2
Lactobacillus bulgaricus	Produces lactic acid from sugars; used for production of yogurt
Salmonella typhi	Causes typhoid fever
Streptomyces griseus	Produces the antibiotic streptomycin
Zymomonas mobilis	Ferments sugar to alcohol and carbon dioxide (for example, in the Mexican drink pulque)

available in the event of breakdowns, and emergency generators can provide power during electric outages. Microbiologists at universities, research institutes, hospital laboratories, and biotechnology companies always maintain experimental and reference microbial strains in their own deep-freezes or in small liquid nitrogen storage tanks. It is ordinarily not feasible, however, for them to have safeguards of the kind available at the ATCC; accordingly, the ATCC recently established a Safe Deposit Service. For a small fee, one can deposit cultures for safe storage, and upon request the ATCC will send you vials of your favorite organism. This service operates like a bank in that only the depositor can make withdrawals.

7

Hardy Survivors in the Microbial Kingdom

Experience with freeze-drying of microorganisms suggests the possibility that cells of certain species may have remained alive in a dormant state for centuries or millenia in natural materials and under natural circumstances. The most plausible candidates for such long-term survivors are bacteria that form *endospores* during the course of their natural growth cycles. Endospores—referred to hereafter simply as spores—are formed mainly by three genera, commonly found in soil:

- *Bacillus*: rod-shaped aerobes
- *Clostridium*: rod-shaped anaerobes (some can fix atmospheric N_2)
- *Thermoactinomyces*: aerobic bacteria that grow best at slightly elevated temperatures (50 °C)

The process of spore formation is essentially the same in these different genera and occurs through a complicated series of events that are under genetic control. Cells growing and multiplying in the presence of adequate nutrient supplies do not form spores. When certain nutrients become exhausted, however, a sequence of changes is triggered, resulting in the formation of a single spore within each "mother" cell. Remnants of the mother cell are eventually sloughed off and a free spore is released.

Spores of bacteria are *not* produced as the result of a sexual process, that is, a process in which two parent cells are involved. Rather, they appear to be life-cycle stages designed to survive "hard times" and to promote dispersal of the species. The most prominent feature of bacterial spores is that they are extraordinarily resistant to adverse environmental conditions. Bacteria that do not form spores are rapidly killed by chemical antiseptics and disinfectants and by relatively mild heating. Bacterial spores, on the other hand, are highly resistant to these and other noxious treatments. The resistance is conferred by the unique architecture of the spore. It has a very low water content (in effect, spores are dehydrated), and the spore wall is a relatively impenetrable shield. Compared with other living cells, bacterial spores are especially remarkable in that they can survive exposure to high temperature, for example, more than 20 minutes in boiling water. Spores of *Clostridium botulinum* can withstand 5 hours of boiling!

Spores are dormant forms that have no detectable metabolic activity. The dormancy can last for long time periods and is ended by certain kinds of environmental triggers, for example, mild heating (to about 70°C, or 158°F) for a few minutes or sudden exposure to particular amino acids. Within minutes, the spore absorbs water, swells, sheds its coat, and develops into the kind of cell that gave rise to the spore in the first place. This series of events is known as germination and outgrowth. Thus, the life history of a spore-forming microbe alternates between a typical cell during "good times" and a spore during times of nutritional stress. Outgrowth followed by cell multi-

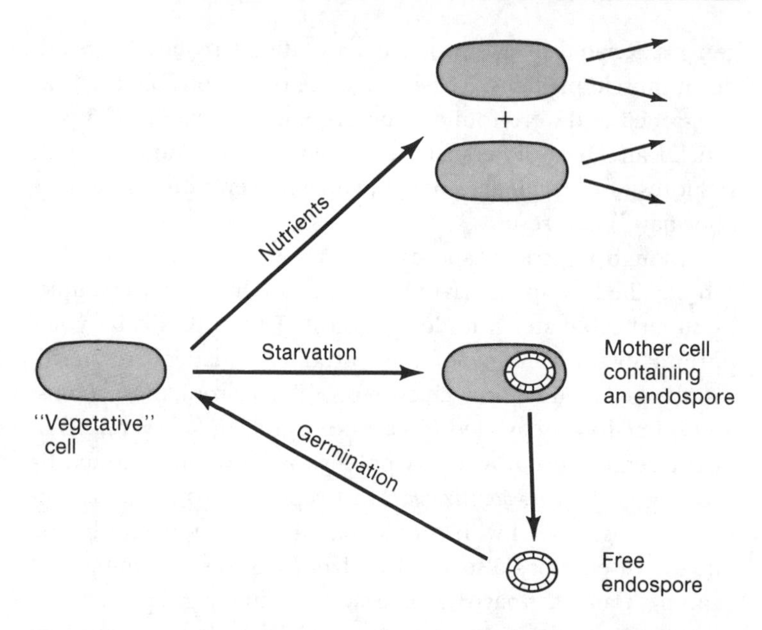

Figure 8 Reproductive modes in bacteria that form endospores. In the presence of required nutrients, the vegetative cell reproduces indefinitely by binary fission. Starvation, however, can trigger the complex process of endospore formation. Under suitable conditions, a free endospore can germinate and "outgrow," yielding a vegetative cell.

plication requires all the nutrients needed for construction of new cells (Figure 8).

SPORE RESISTANCE AND PUBLIC HEALTH

Various species of clostridia bacteria are normal inhabitants of the intestinal tract of humans and animals. Consequently, manured soils contain many spores of *Clostridium perfringens* and its close relatives. If these are accidently introduced into a deep wound where conditions become anaerobic and nutrients are available, the spores can germinate and grow. Growing cells of *C. perfringens* rapidly ferment the sugars present in muscle tissue, giving rise to large amounts of carbon dioxide and hydrogen gas. Related species produce devastating enzymes that break down muscle proteins. These bacteria are responsible for "gas gangrene" and were found to be present in more than 75 percent of affected soldiers studied during World Wars I and II. If spores of *C. tetani* are contaminants in deep wounds, tetanus ("lockjaw") can result.

Although clostridial spores are very resistant to heat, they can be killed by appropriate heating procedures—for example, by superheated steam under pressure. This is particularly important in the canning of food products. Inadequate sterilization of canned salmon, chicken broth, macaroni, mushrooms, and other foods have led to cases of botulism, a severe, often fatal disease caused by a very potent toxic protein secreted by growing cells of *C. botulinum*. Most reported instances of botulism are associated with contaminated foods canned at home, but the disease has also resulted from failures in commercial canning. Cans that have become swollen (due to gas production by clostridia) or whose contents smell foul (due to metabolic products of clostridia) are suspect. In the United States during the 1980s, the Food and Drug Administration ordered recall of millions of cans of foods because there was reason to believe some of them might have been contaminated with *C. botulinum* spores.

CRYPTOBIOSIS: "LATENT LIFE"

During the 1920s and 1930s, a series of technical publications claimed demonstration of "amazing longevity" of microbes within very ancient natural materials such as coal and meteorites. Similarly, in the 1960s reports appeared describing the presence of ancient viable bacteria in salt deposits laid down millions of years ago. Experienced microbiologists have discounted these claims on several grounds. The experimental techniques used for such studies must be flawless to avoid contamination with contemporary microbes from our surroundings. It is also clear that the possibility of seepage of modern microbes into ancient natural materials over the course of decades or centuries must be conclusively eliminated. These criteria were not satisfied in the studies noted.* The question still remains—what is the maximal longevity of microbes under favorable natural conditions?

Cryptobiosis, or "latent life," is defined as the state of an organism when it shows no visible signs of life and when its metabolic activity becomes hardly measurable, or comes (reversibly) to a standstill. This phenomenon has fascinated scientists for almost three centuries, starting with the experiments of Leeuwenhoek. In 1745, Henry Baker, a distinguished English naturalist, confirmed Leeuwenhoek's observation that tiny "wheel animals" (rotifers) remained alive for surprisingly long periods in the dry state. This was described in a book published by Baker in 1753 (*Employment for the Microscope*, R. Dodsley, London):

> Mr. Leeuwenhoek kept some Dirt, taken out of a Leaden Gutter, and dried as hard as Clay, for twenty-one Months together; and yet when it was infused in Water, Multitudes of these Creatures soon appeared unfolding themselves, and quickly after began to put out their Wheels; and I myself have experienced the same with some that had been kept much longer. . . .I cannot conclude this Subject without doing all the Honour I am able to the Memory of Mr. Leeuwenhoek, by repeating that we are obliged to his indefatigable Industry for the first Discovery of this most surprising Insect.

The problem of long-term viability of plant seeds has attracted the attention and experimental efforts of agriculturalists

---

*In 1931, C. B. Lipman published an account of experiments that supposedly demonstrated the existence of living bacteria in coal from both Wales and Pennsylvania. This was followed by reports of similar experiments; for example, see Lipman, C. B., 1934, Further evidence on the amazing longevity of bacteria, *Science* 79:230. However, others investigated Lipman's claims and concluded it was possible that Lipman's samples were contaminated with bacteria of recent origin (Burke, V., and Wiley, A. J., 1937, Bacteria in coal. *Journal of Bacteriology* 34:475).

and other scientists for some time. In 1850, it was demonstrated that 85 percent of lotus seeds from a 150-year-old collection in the British Museum retained their viability, that is, they could be germinated after 150 years, yielding normal plants. A number of reports on viability of presumably still older seeds presents the problem of accurate determination of their ages. This can be done using the carbon-14 dating technique, and it now seems that the record (authenticated) longevity of lotus seeds is about 700 years.* Such longevity is indeed remarkable. Since bacterial spores appear to be more resistant to adverse environmental circumstances than other life forms, we can anticipate that spores of microbes might have even greater longevity.

GREAT LONGEVITY OF MICROBES

A number of studies have shown that spores of *Bacillus anthracis* and *Clostridium tetani* can remain viable for at least 50 to 70 years. For example, anthrax spores prepared by Pasteur in 1888 were found to be alive 68 years later. As of 1962, the longest survival recorded was 118 years for a strain of bacillus in an old can of meat. The time range was significantly extended by a classic study by P. H. A. Sneath.† He realized that British botanists had been collecting plant specimens in a systematic way since about 1640. These were dried, packaged, and stored in the Herbarium of Kew Botanic Garden near London. Sneath obtained samples of dry soil adhering to the roots of these specimens and examined them for the presence of living bacteria. His experiments clearly showed viable *Bacillus* spores in samples as old as 320 years. Moreover, from his results he could calculate the death rate during storage, and from this he esti-

*The seeds were collected from silt below the surface of an ancient dried pond in a Chinese village. Beware of enterprising guides in ancient tombs who sell samples of modern grain under the guise of "mummy wheat" or "mummy barley."

†See Sneath, P. H. A., 1962, Longevity of micro-organisms. *Nature* 195:643.

mated that a ton of dry soil would still contain a few viable *Bacillus* spores even after 1000 years.

A few years after Sneath's research, evidence was obtained for long-term survival of spore-forming bacilli in sediments below the Pacific Ocean floor (off the coast of Mexico and southern California). Sediment at a depth of 150 centimeters was determined to be at least 5800 years old and contained 25 to 75 viable cells of *Bacillus* per gram of wet sediment. It seemed that bacilli could survive, presumably in the form of spores, for thousands of years in cold, dark, wet environments. Control experiments clearly indicated that the viable cells were not contemporary microbe contaminants introduced during retrieval of the sediment from the ocean depths. Nevertheless, the investigators* felt obliged to take a conservative stance:

> It is possible that 'alien' spore-forming bacteria may have persisted in such sediments for these long periods of time. However, the burden of proof would fall on anyone who attempted to make such a statement, since this would imply bacterial spore ages of many thousands of years.

Bacteria of the genus *Thermoactinomyces* produce heat-resistant spores and grow optimally at a temperature of about 50°C (122°F). They are present in most soil samples and sporulate profusely in habitats such as compost, haystacks, and stored cereals. Research during the past decade has provided convincing evidence for survival of *Thermoactinomyces* spores for time periods of 1900 to 2700 years in sediments deposited under lakes of the English Lake District and in an ancient lake bed in East Anglia. Viable *Thermoactinomyces* spores were also found in occupational debris from a Roman archeological site at Vindolanda, North Umberland (United Kingdom).† One stra-

---

*From Bartholomew, J. W., and Paik, G., 1966, Isolation and identification of obligate thermophilic spore-forming bacilli from ocean basin cores. *Journal of Bacteriology* 92:635.

†Seward, M. R. D., Cross, T., and Unsworth, B. A., 1976, Viable bacterial spores recovered from an archaeological excavation. *Nature* 261:407.

tum of the debris, dated between A.D. 85 and 95, contained at least 4000 living spores per gram of material. The occupational debris, rich in organic litter (bracken, straw, etc.) was sandwiched between compacted layers of clay, and preservation of spore viability was probably enhanced by anaerobic and other favorable chemical and physical conditions.

Additional evidence for remarkable longevity of *Thermoactinomyces* spores comes from a study of a lake bed in Minnesota (Elk Lake)* A 20-meter-deep core taken of the lake bed deposits (collected in 1978) shows a distinct record of annual laminations (layers of sediment) going back more than 10,000 years. Viable *Thermoactinomyces* spores were detected in various layers, including sediments deposited about 7000 to 7500 years ago. The possibility that the recovered cells were actually younger spore contaminants that were somehow displaced downward through older sediment is considered unlikely.

The accumulation of evidence for viability of bacterial spores for thousands of years can no longer be attributed simply to artifactual observations and gives encouragement for renewed efforts in the field of archaeomicrobiology. With the development of new and more sensitive techniques of dating and molecular biological analysis, the study of ancient spores may contribute to a better understanding of how bacterial species evolve over time spans of tens of thousands of years. The study of the stability of spores over short time courses may aid in this endeavor. An example is provided by spores of the causative agent of anthrax, *Bacillus anthracis*, which are thought to survive in soil for significant time periods. This has been confirmed by studies on so-called anthrax island. During World War II, trials of *B. anthracis* as a potential agent of biological warfare were carried out on Gruinard Island, which lies off the west coast of

---

*Parduhn, N. L., and Watterson, J. R., 1985, Recovery of viable *Thermoactinomyces vulgaris* and other aerobic heterotrophic thermophiles from a varved sequence of ancient lake sediment, *in Planetary Ecology*, Caldwell, D. E., Brierley, J. A., and Brierley, C. L., eds, Van Nostrand–Reinhold, Princeton, N. J., p. 41.

Scotland. Tethered sheep were exposed to large clouds of *B. anthracis* spores released by detonation of small bombs. As a result, the island was heavily contaminated. A survey in 1979 disclosed the persistence of virulent spores in the soil. Although Gruinard Island is off-limits to the public, foreign tourists unable to read the warning signs occasionally stray onto the island. Attempts are being made to rid the island of anthrax spores by treatment with appropriate chemicals.

'GRUINARD ISLAND'
THIS ISLAND IS
GOVERNMENT PROPERTY
UNDER EXPERIMENT.
THE GROUND IS CONTAMINATED
WITH ANTHRAX AND DANGEROUS.
LANDING IS PROHIBITED.
BY ORDER ·1981·

Gruinard Island has been posted with signs that warn of the presence of viable spores of the "anthrax bacillus."

8

Microbes and the Carbon Cycle

Microbes are prominent agents in the recycling of several major chemical elements on Earth, notably oxygen, carbon, nitrogen, and sulfur. Element recycling involves sequential conversion of one form of an element to other forms and eventual reconversion to the original state. If recycling of the elements noted were to stop suddenly, all forms of life would soon come to an end. A large diversity of microbes participate in this global chemistry, and generations of microbiologists have been kept busy investigating this aspect of the microbial universe. This research has disclosed an amazing spectrum of lifestyles, that is, capacities to grow in a great variety of nutritional circumstances. Microbes occur almost everywhere on Earth, even in ecological niches that would be considered extreme or uninhabitable for other forms of life.

If all microbes were suddenly to die because of some great natural catastrophe, all life on Earth would eventually cease. Some years ago an imaginative microbiologist graphically described the unhappy sequence of events that would ensue if the

Earth were to collide with the tail of a comet containing a mysterious gas that could destroy all microbes without doing any damage to plants or animals.* After the first sigh of relief in anticipation of a future free of certain contagious diseases, we would be faced with such ramifying difficulties as a diminishing CO_2 (carbon dioxide) content in the atmosphere, followed by decreasing plant life, then no milk (cows live mainly on grass), then unknown diseases due to lack of vitamins normally produced in our intestines by helpful bacteria, and persistent sewage in our water supplies. Then the trouble would really begin—we would be smothered by the organic excretions of animals and the accumulated debris of dead plants and animals.

THE CARBON CYCLE

This dreadful scenario reflects the roles that microbes play in the recirculation of carbon atoms on the Earth, otherwise known as the "carbon cycle." The breakdown of the organic components of dead animals and plants is accomplished by a myriad of microbes present in soil and all other natural environments. The microbes decompose organic matter to obtain energy and/or nutrients for their own multiplication in several ways including fermentation. Numerous species of microbes are engaged in this phase of the carbon cycle, which results in the conversion of organic carbon to CO_2. This overall process occurs on a gigantic scale. Let's now examine how carbon moves through the cycle.

In contrast to animals which require organic compounds of carbon, plants grow on CO_2, the major form of inorganic carbon. The utilization of CO_2 by green plants through photosynthesis (the process of using light to convert CO_2 to organic

---

*From Rahn, O., 1945, *Microbes of Merit.* Jaques Cattell Press, Lancaster, PA.

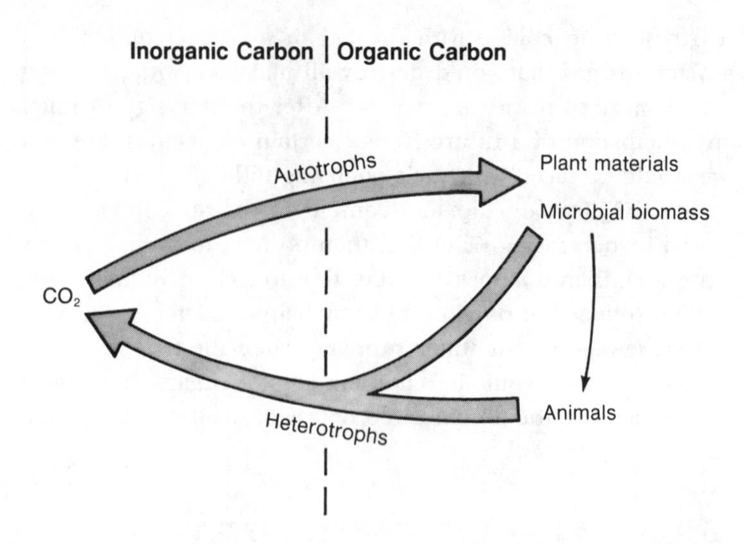

Figure 9 The carbon cycle. Carbon atoms are in constant flux between inorganic and organic forms. Autotrophs convert CO_2 to a multitude of organic compounds, and these are eventually recycled back to CO_2 by the chemical activities of diverse assortments of microbes.

carbon) is the largest chemical process on Earth. It has been estimated that about 300 billion tons of organic carbon are produced on the Earth each year from CO_2; this is roughly 100 times the total output of the chemical, metallurgical, and mining industries on our planet. The annual rate of CO_2 utilization by plants is such that all of the CO_2 in our atmosphere would be exhausted in about 30 years if it were not constantly replenished. Obviously, CO_2 is being continually regenerated on a vast scale. This is another way of saying that carbon atoms on the Earth must "flow" in a cycle between inorganic and organic forms. Microbes are important agents of much of this carbon atom traffic, which is illustrated in Figure 9. The diagram indicates only the general outline of the carbon cycle, and introduces two new terms:

- *autotroph*: an organism that uses CO_2 as its sole or primary source of cellular carbon

Table 5 Examples of Autotrophic Bacteria That Can Convert CO_2 to Organic Compounds

Organism	Relationship to O_2	Special Attributes
Chromatium vinosum	Anaerobe	Uses light as energy source (but does not produce O_2)
Methanobacterium thermoautotrophicum	Anaerobe	Requires H_2 gas; produces methane; can grow at high temperature
Thiobacillus thiooxidans	Aerobe	Uses sulfur as the source of growth energy; can grow in very acidic media
Nitrosomonas europaea	Aerobe	Uses ammonia (NH_3) as the source of growth energy
Nostoc muscorum	Aerobe	Uses light as energy source (and produces O_2); also can use N_2 as the nitrogen source for growth

- *heterotroph*: an organism that requires organic compounds as sources of cellular carbon and energy

Many of the processes that are represented by the broad arrows in Figure 9 are anaerobic, whereas others require the participation of O_2. The major part of the carbon atom flow between CO_2 and organic carbon is accomplished by plants and cyanobacteria (see Chapter 13) that carry out a photosynthetic process that produces oxygen. Examples of different kinds of autotrophic microbes are given in Table 5. Surprisingly, it has been estimated that as much as 80 to 90 percent of the annual flow in this direction occurs in the oceans. As noted earlier, flow of carbon atoms in the reverse direction (often referred to as mineralization) is catalyzed by heterotrophic microbes. This occurs in sequential stages: different species with particular capabilities come into play at different points in the gradual decomposition of organic biomass to CO_2.

THE ROLE OF PHOTOSYNTHESIS

Green plant photosynthesis not only provides organic carbon for animal life, but also generates O_2, which is essential for the energy metabolism of all animals. The interdependence of plant and animal life was first discovered in 1772 by the Englishman Joseph Priestley (1733–1804), an ordained minister and brilliant scientist whose wide interests included electricity, optics, and gases. His experiments represent one of the truly great moments in the history of biological science. The following quotations are from his celebrated book *Experiments and Observations on Different Kinds of Air* (printed for J. Johnson, London, 1774).

> I have been so happy as by accident to hit upon a method of restoring air which has been injured by the burning of candles, and to have discovered at least one of the restoratives which nature employs for this purpose. It is *vegetation*. . . .One might have imagined that, since common air is necessary to vegetable, as well as to animal life, both plants and animals had affected it in the same manner; and I own I had that expectation when I first put a sprig of mint into a glass jar standing inverted in a vessel of water: but when it had continued growing there for some months, I found that the air would neither extinguish a candle, nor was it at all inconvenient to a mouse, which I put into it. . . .
>
> Finding that candles burn very well in air in which plants had grown a long time. . .I thought it was possible that the same process might also restore the air that had been injured by the burning of candles. Accordingly, on the 17th of August 1771, I put a sprig of mint into a quantity of air in which a wax candle had burned out, and found that on the 27th of the same month, another candle burnt perfectly well in it. This experiment I repeated, without the least variation in the event, not less than eight or ten times in the remainder of the summer.

This classic experiment is illustrated in Figure 10 (also see Figure 11). Priestley discovered O_2 gas (which he called "dephlogisticated air") and demonstrated that O_2 is a vital link

Figure 10 Classic Priestley experiment showing interdependence of animal and plant life. A lone plant and a lone mouse in separate closed jars died, but when a plant and mouse were placed together in a closed jar they continued to live.

between animal and plant life. Priestley was a minister who eventually became Preacher at the New Meeting House in Birmingham, England. A nonconformist and theological dissenter, he was also politically active and supported early phases of the French Revolution. These activities led a "Church-and-King" mob to destroy the New Meeting House as well as Priestley's

Figure 11 Apparatus used by Priestley. For his early experiments on gases, Priestley frequently used household utensils (wine and beer glasses, clay tobacco pipes, a laundry tub, etc.). Later, Josiah Wedgwood supplied him with ceramic tubes, dishes, crucibles, and other items.

home and laboratory in 1791. Continued political persecution prompted him to emigrate to Northumberland, Pennsylvania in 1794. Priestley's numerous publications dealt with a wide range of subjects that included language, psychology, politics, and theology in addition to his scientific studies. His memorial states, "His discoveries as a philosopher will never cease to be remembered and admired by the ablest improvers of Science."

The production of the organic matter of plants from CO_2 and water requires a large input of energy, and in photosynthesis, this is provided by light. The aerobic respiration of animals (and other heterotrophs) provides their energy needs, and from the standpoint of energy, it is the reverse of photosyn-

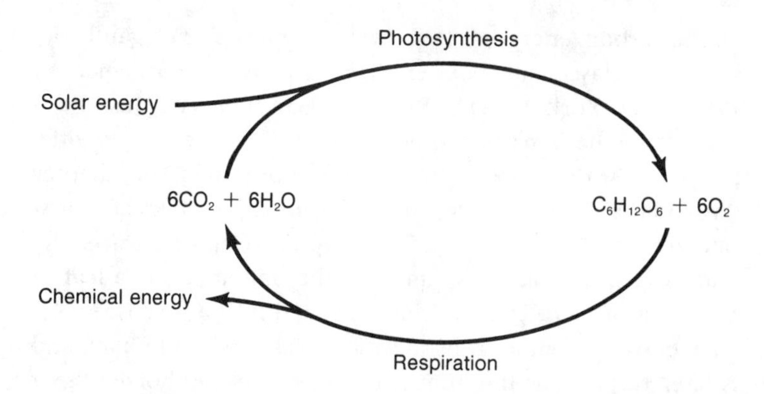

Figure 12 This diagram summarizes the two major biological processes which support all plant, animal, and microbial life on Earth. Solar energy drives the conversion of CO_2 and water to organic matter, symbolized here as sugar ($C_6H_{12}O_6$) in plants. Oxygen gas (O_2) is also produced, and is the source of the O_2 required by animals and many aerobic microbes (some bacteria use a form of photosynthesis that does not yield O_2; see Chapter 13). In the process of respiration, animals and aerobic microbes "burn" sugar and other organic compounds with O_2 to obtain chemical energy (ATP) for growth and metabolism.

thesis. Thus, the solar energy locked into sugar molecules by photosynthesis is made available by respiration. The carbon/energy cycle can be summarized as shown in Figure 12. About 16 times every minute, every human inhales, sucking air into the lungs, and then exhales. The air taken in contains about 21 percent O_2 and that exhaled about 16 percent. The missing 5 percent is absorbed during the time the air is in the lungs and is used in various tissues for the respiration process. Exhaled air is enriched in the products of respiration—CO_2 and water. On a cold day, condensation of water vapor in exhaled air into small liquid droplets makes the breath visible. As in photosynthesis, respiration is effected by a complex series of reactions involving numerous enzyme catalysts. Some of these respiratory catalysts are specifically and strongly inhibited by low concentrations of cyanide and carbon monoxide, which explains the poisonous actions of these compounds.

The carbon/energy cycle shown in Figure 12 is elegantly described in a fascinating book entitled *The Periodic Table* (Schocken Books, New York, 1984) by Primo Levi, an Italian chemist. Each chapter of the book bears the name of an element as its title, and deals with episodes from the author's life. The chapter "Vanadium," for example, chronicles his experiences as a prisoner/chemist in Auschwitz. The chapter "Carbon" traces the path of a single carbon atom from the time it enters a leaf in the form of CO_2, becomes incorporated into a glucose molecule, ends up (temporarily) in wine consumed by a human, and is later respired as the human pursues a bolting horse. "So a new molecule of carbon dioxide returned to the atmosphere, and a parcel of the energy that the sun had handed to the vine-shoot passed from the state of chemical energy to that of mechanical energy, and thereafter settled down in the slothful condition of heat, warming up imperceptibly the air moved by the running and the blood of the runner. . . . Every two hundred years, every atom of carbon that is not congealed in materials by now stable (such as, precisely, limestone, or coal, or diamond, or certain plastics) enters and reenters the cycle of life, through the narrow door of photosynthesis."

COAL AND OIL

In concluding discussion of the carbon cycle, it is pertinent to note an important geological abnormality. During certain periods of the Earth's history, some organic carbon of dead plant material was not directly recirculated. This was particularly true about 250 to 300 million years ago (Pennsylvanian Period). At that time, large quantities of plants in swamps and lagoons were covered by deposits of muds carried by invading seas and rivers. Shut off from sunlight, the plants died, and their decomposition was greatly slowed due to shortage of O_2 in the muds. The organic matter was consequently converted to peat (partly decayed vegetation), some of which was further transformed to coal. In other words, coal deposits represent huge amounts of modified organic plant materials that have escaped the dynamic carbon cycle. When we burn coal and thereby convert it to CO_2, we are accelerating return of carbon to the active cycle. Combustion of oil also restores CO_2 to the atmosphere, and there is considerable evidence indicating that oil deposits were formed, in part, by ancient microbial processes.

Energy: A Villanelle*

The log gives back, in burning, solar fire
green leaves imbibed and processed one by one;
nothing is lost but, still, the cost grows higher.

The ocean's tons of tide, to turn, require
no more than time and moon; it's cosmic fun.
The log gives back, in burning, solar fire.

All microörganisms must expire
and quite a few became petroleum;
nothing is lost but, still, the cost grows higher.

The oil rigs in Bahrain imply a buyer
who counts no cost, when all is said and done.
The logs give back, in burning, solar fire.

but Good Gulf gives it faster; every tire
is by the fiery heavens lightly spun.
Nothing is lost but, still, the cost grows higher.

So guzzle gas, the leaden night draws nigher
when cinders mark where stood the blazing sun.
The logs give back, in burning, solar fire;
nothing is lost but, still, the cost grows higher.

— *John Updike*

---

*From *Facing Nature* by John Updike. Copyright © 1985 by John Updike. Reprinted by permission of Alfred A. Knopf and André Deutsch Ltd. Originally published in *The New Yorker*.

9

Bacteria That Produce and Use Methane

DISCOVERY OF NATURAL METHANE FORMATION

The physicist Allesandro Volta (1745–1827) is best known for his discoveries on the nature of electricity and his name was truncated to a household word—*volt*, the basic unit of electrical force. Volta is particularly famous for his invention of what was called the "voltaic pile," now known as the electric battery. In 1801, he was invited to demonstrate the device to Napoleon, who was entranced. Napoleon bestowed the title "Count and Senator of the Kingdom of Lombardy" on Volta, and also a generous lifetime pension.

Volta was also interested in other physical phenomena. In 1776 Volta visited Lake Maggiori where he noticed bubbles rising to the surface of the water, especially in shallower and marshier locations. He collected some of the gas, and using his "electric pistol," which was a revolver-like device he invented for igniting combustible gases in closed vessels, he found that it was flammable. Volta concluded that the flammable "marsh gas" originated from decaying organic matter and, indeed, he was correct. Thirty years later, Volta's combustible gas was identified as methane (CH_4), and after a lapse of another sixty years, one of Pasteur's students obtained the first evidence (1868) indicating that methane is produced by microbes.

We now know that the formation of CH_4 is actually the last phase of the anaerobic decomposition of photosynthetically produced organic matter by a large assortment of microbes with different appetites (in other words, some take over where others leave off). The final microbes in this so-called food chain are the *methanogens*, a group of very anaerobic bacteria that generate methane. (A new species discovered in 1970 was recently named *Methanococcus voltae* after Volta.) Pure strains of methanogens in the laboratory tend to be very sensitive; they will not grow unless every trace of oxygen is first removed from the nutrient medium. Thus, it might be difficult to imagine where methanogens occur in nature. There is a surprising diversity of suitable anaerobic habitats. Obviously, one place is in deposits of mud, which accounts for Volta's observations. The presence of flammable methane inside living trees was noted as early as 1907; this occurs in poplar and certain other trees that grow on poorly drained soils near lakes and rivers. If a hollow metal tube is drilled into a tree of this sort (which contains a pulpy anaerobic midsection called wetwood), the gas that escapes can usually be ignited producing a blue flame. Recently, the methanogenic bacteria responsible have been isolated and characterized. Methanogens are also present in the intestinal contents of all animals, including humans, and consequently they always abound in sewage.

METHANOGENS IN RUMEN SYMBIOSIS

The term *rumen symbiosis* is used to refer to the complex interplay between a ruminant animal, such as the cow, and the microbes present in its prestomach, the rumen. In a typical cow, the rumen contains as much as 100 liters (100,000 milliliters) of fluid that is teeming with single-cell animals (protozoa) and bacteria of numerous kinds. Each milliliter of the fluid contains about one million protozoa and about ten billion bacteria. The rumen "incubator" is a kind of biological–chemical factory in which microbes produce the actual energy nutrients of the cow, primarily from cellulose and from other organic matter in grass and fodder. Microbes in the cow's rumen fluid break down cellulose to simple sugar units, and these are then fermented to an assortment of even simpler energy-rich products. The cow gets its energy by "burning" (respiring) these microbial products in its own tissues. Of special interest here is the fact that the gas composition of the rumen consists of about 40 percent methane and 60 percent carbon dioxide. The cow must eliminate gases by belching, otherwise it comes down with an ailment called "bloat." A 1000-pound cow produces about 200 liters of methane per day. In a whimsical moment Professor Rodney Quayle of the University of Bath calculated that: ". . .the cattle population (1967) of the United Kingdom eructating in concert, could have filled the airship 'Hindenburg' with methane in 114 minutes." But there are easier ways to collect the CH_4 resulting from anaerobic decomposition of organic matter.

It was noted above that the rumen contains large amounts of CO_2. The gas is generated by microbial breakdown of organic substances consumed in the diet, and is converted to methane by a "hydrogenation" process unique to so-called methanogens. In this process, four molecules of hydrogen gas (H_2) are used to convert one molecule of CO_2 to one of CH_4:

$$CO_2 + 4H_2 \rightarrow CH_4 + 2H_2O + \text{energy (ATP)}$$

The mechanism is quite complex, involving numerous enzymes, and yields energy for growth of methanogens. Such methanogenic bacteria are similar to green plants in that they also can grow on CO_2 as the only carbon source for making all cellular substances (in other words, they are autotrophs). In the methanogens, however, only a small fraction of the available CO_2 is converted to cell materials; a larger fraction of the CO_2 is hydrogenated, yielding methane.

How do methanogens obtain hydrogen gas? In laboratory experiments, the fermentation of organic substances by species of bacteria found in the rumen (and in the intestinal contents of humans) is always characterized by the production of H_2. This also occurs naturally in the rumen, but the methanogens present here are so active in using the H_2 that its concentration in the rumen atmosphere always remains extremely low. Hydrogen gas acts as a connecting link between anaerobes that are obliged to make H_2 for their own energy supply as they decompose organic substances and the methanogens which require the H_2 for their energy. This is spoken of as "interspecies H_2 transfer" and is believed to be of importance for the ecology of many anaerobic microbes.

METHANE PRODUCTION IN LANDFILLS

"Landfills Spread Methane, Fear in Virginia City" was a headline in the *Louisville Courier-Journal* on April 5, 1976, inspired by the plight of 1000 families in Richmond, Virginia whose well-being and homes were threatened by methane seeping from nearby landfills. Methane is colorless, odorless, combustible, and highly explosive in the presence of oxygen gas. The newspaper further explained:

> The methane problem is relatively unknown because sanitary landfills, which produce the gas, are comparatively new in America. They were first introduced after World War II, and it is just now, the officials believe, that trash in the landfills is beginning to decay on a massive scale. It is underground decay, in the absence of oxygen, that produces methane—an extremely volatile gas responsible for most coal mine explosions and also known as 'swamp gas.' It can seep out of landfills along sewer and other pipes or through underground fissures, endangering nearby houses. Richmond officials discovered the danger January 8, 1975, when a woman in an apartment alongside a city landfill saw a

> blue flame dart from a bedside lamp across the room and explode in her living room. Fire Chief John F. Finnegan said that 'it is amazing' that neither the woman nor her husband was more seriously injured in the explosion, which blew out the apartment door and two windows. The woman suffered first-degree burns of the hand and her husband's hair was singed, Finnegan said.

The circumstances in landfills, rich in cellulosic and organic wastes, are basically the same as in the rumen, but obviously not nearly as well controlled. There is one important initial difference: in the landfill, oxygen (in air) is present for a while, but it is gradually removed by the metabolic activities of aerobic microbes and when anaerobic conditions prevail the methanogens begin to grow.

METHANE FORMATION IN ANAEROBIC DIGESTERS

Waste cellulose and other organic materials are decomposed to methane in still another type of anaerobic locale, the sewage sludge digester (see Chapter 15). Anaerobic digestion by microbes is an essential phase of sewage waste treatment. In the digester container, "bio-gas" consisting of CH_4 and CO_2 is generated by the microorganisms present in raw sewage. The methane produced by these digesters provides a simple and economical form of usable energy. In most large sewage works the bio-gas is used to power diesel engines that operate pumps and generators of the works. Digesters usually operate at 35 to 40°C (95 to 104°F), and this temperature is maintained by burning the bio-gas in special boilers.

Simple anaerobic digesters can be easily constructed and fed with a variety of organic wastes, such as cowdung. The basic design consists simply of a fermentation pit, acting as the digester, and a gasholder of some sort that floats over the pit; various construction materials can be used, such as reinforced concrete. Since they are so simple, small bio-gas generators seem unimpressive, but they can be rapidly made and are particularly useful for satisfying the energy needs of small groups of people. Calculations of actual agricultural energy budgets for farms in the western world show that the methane gas generated from the manure of a herd of 100 to 200 cattle could generate sufficient gas to satisfy all the farmstead heating requirements.

An important example of the promise of bio-gas is provided by recent developments in India and China. The energy crisis in India during the late 1970s led the Indian government to promote construction of bio-gas digesters in which cowdung is fermented. A planning commission estimated that India has a potential of at least 18,750,000 family-sized bio-gas plants and 560,000 community-sized plants. According to a report in the scientific journal *Nature**:

> If this potential were realized by 1990, bio-gas could supply India with an energy equivalent to nearly 44 percent of its projected electricity consumption, and reduce its projected consumption of coal by 15 percent and firewood by 79 percent. The use of firewood would be reduced to an environmentally safe level, saving forestry expenditure, and the organic manure by-product would also reduce expenditures on chemical fertilizers.

Research is proceeding in India on use of agricultural wastes (such as straw from wheat or rice) as another type of feedstock for bio-gas production. The *Nature* article also stated that China is reported to have set up more than seven million bio-gas plants in a few years.

TERMITES AS A SOURCE OF METHANE

In 1982, new studies indicated that termites represent a hitherto unsuspected source of methane pollution of the Earth's atmosphere, continued pollution of the kind that could even-

---

*Can bio-gas provide energy for India's rural poor. *Nature*, September 6, 1979, pp. 9–10.

tually affect weather patterns. The digestive tracts of termites contain large numbers of methanogens, other anaerobic bacteria, and protozoa; these microbes efficiently process great quantities of wood and other biomass. It is estimated that for every person on Earth, there is 3/4 of a ton of termites! Calculations indicate that bovine flatulence adds 85 million tons of methane to the atmosphere annually, and revised estimates for termite methane production suggest that it may be of the same order of magnitude.

THE ENERGY VALUE OF METHANE

For assessing the utility of energy-rich organic materials as fuels, the so-called *heat of combustion* is a useful quantity. This is determined by burning a weighed amount of the material in a calorimeter, an apparatus that measures the amount of evolved heat. This heat can be expressed in various ways, one of the simplest ways being kilocalories per kilogram. One calorie is the amount of heat needed to raise the temperature of 1 gram of water from 15 to 16°C. Correspondingly, one kilocalorie is needed to raise the temperature of 1000 grams (1 kilogram) of water from 15 to 16°C. We are interested in determining the number of kilocalories of heat liberated per kilogram of the substance burned. Heats of combustion of special interest to us here are given in Table 6. As shown in the table, the heat of combustion of methane substantially exceeds that of alcohol, and that of molecular hydrogen (H_2) is still greater. Molecular hydrogen has been discussed in recent years as a potential fuel, and numerous schemes have been suggested for production of H_2 in massive amounts. In principle, large quantities of H_2 could be made from water by a combination of physical and chemical processes. Alternatively, there are possibilities for exploiting biological systems that evolve H_2. On the basis of information now available, however, the prospects for development of a dependable biological process that could operate on a meaningfully large scale are doubtful.

Table 6 Heats of Combustion of Various Substances

Substance	Heat of Combustion (Kilocalories per Kilogram)
Glucose	3,735
Ethyl alcohol	7,140
Benzene	10,000
Gasoline	10,500
Propane	11,980
Methane	13,200
Hydrogen gas	34,200

Eventually, oil will become scarce and more costly. Methane is consequently attracting renewed attention as a motor fuel for the future. To quote from an expert*:

> The earth's crust contains large amounts of methane. The gas can also be obtained from biomass and from synthetic gas derived from coal. In the United States, a million-mile pipeline network exists for distribution of the gas. Methane is already being used in about 400,000 vehicles around the world, including 250,000 in Italy and 20,000 to 30,000 in the United States. Users have found that engine wear is reduced: lubricating oil is not diluted as it is when gasoline is used. Exhaust gases are relatively non-polluting. Start-up of motors is not affected by cold weather. An engine designed especially for methane has an energy efficiency greater than that of ordinary automobiles.

METHYLOTROPHS

Microbes that can use methane or methyl alcohol (CH_3OH) as their sole source of carbon and energy are widely distributed in nature (in mud, natural waters, and soils). Organisms with this capacity are called *methylotrophs* because the common chem-

*Abelson, P. H., 1982, Methane: A motor fuel. *Science*, 218:641.

ical feature of methane and methyl alcohol is the methyl group, written chemically as

```
    H
    |
H—C—
    |
    H
```

Methane utilization is restricted to certain bacterial species, which frequently can also grow on methyl alcohol. In contrast, certain eucaryotic yeasts grow well on methyl alcohol but cannot use methane.

Aside from the fact that methylotrophs play roles in the Earth's carbon cycle, they are of interest for biotechnological applications, especially in regard to producing protein food supplements. By the end of this century, the Earth's human population is expected to increase at least 50 percent, to a total of about 6 billion. Many experts believe that the greatly increased demands for protein in human nutrition will not be easily met by improvements in agricultural practices. Accordingly, methylotrophs are being investigated as alternative sources of protein for humans. Methane (the major component of natural gas) is comparatively cheap and is also easily converted to methyl alcohol. Thus, the methylotrophs are attractive for commercial exploitation. For microbial cells to be useful as protein food supplements, several criteria must be met. The cells must

1. be digestible and have an acceptable taste;
2. be free of harmful substances;
3. have a relatively high protein content (40 to 75 percent); and
4. have a comparatively low content of nucleic acids. (High levels of nucleic acids in human diets can lead to formation of stones in the urinary tract, and gout.)

Obviously, the growth of the cells on a large scale must also be economically feasible. On the whole, methylotrophs meet the criteria noted, and industrial-scale production of cells grown on methane or methyl alcohol has already been initiated by oil companies and other commercial concerns. Such efforts are usually referred to as "single-cell protein" production, meaning that a single type of microbial cell is being grown. Imperial Chemical Industries in England has marketed dried cells of methylotrophic bacteria, grown on methyl alcohol, under the name "Pruteen." The latter contains 16 percent nucleic acids, too high for human consumption, but domestic animals, especially poultry, do well on it. We can expect that continued research on single-cell protein production will result in a number of useful food supplements for humans.

10

Microbes Recycle Nitrogen

The carbon cycle is intertwined with another major element cycle, namely, that of nitrogen (Figure 13). When carbon dioxide (CO_2) is converted to cellular organic matter by autotrophs, inorganic nitrogen is incorporated into the structures of organic biomolecules, principally proteins and nucleic acids. The forms of inorganic nitrogen of greatest relevance here are atmospheric nitrogen gas (N_2), ammonia (NH_3), and nitrate (NO_3). Nitrate is always associated with a mineral element such as potassium, sodium, calcium, or magnesium.

When the organic matter of dead organisms is mineralized by the actions of heterotrophic microbes, the nitrogen of proteins and nucleic acids is released in the form of ammonia. This is known as "ammonification" and is indicated as one phase of the nitrogen cycle depicted in Figure 13. After ammonia begins to accumulate in organic matter rich in nitrogen, certain other species of aerobic bacteria that can use ammonia as an energy source begin to flourish—these are the autotrophic "nitrifying bacteria" (such as *Nitrosomonas europaea*; see Table 5). The even-

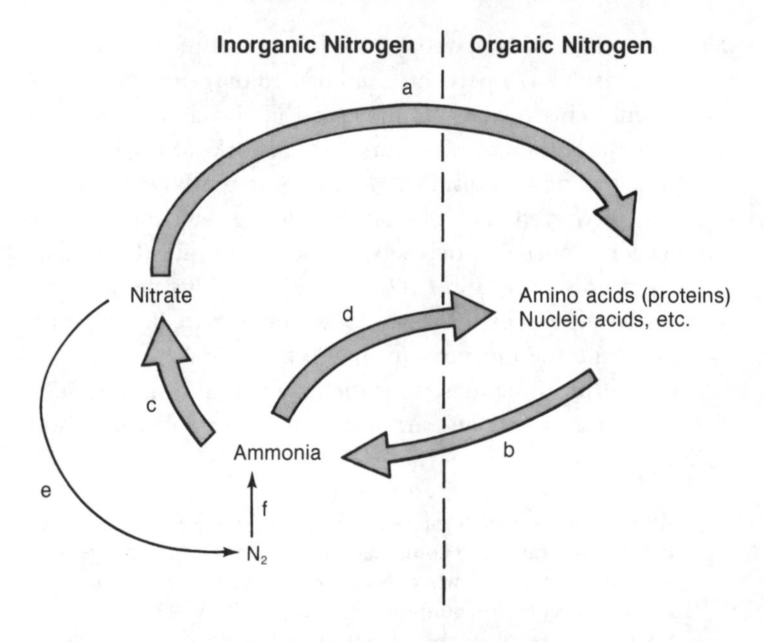

Figure 13 The nitrogen cycle. (a) Nitrate assimilation (→ plants → animals), (b) Ammonification, (c) Nitrification, (d) Ammonia assimilation, (e) Denitrification, and (f) Nitrogen fixation.

tual product is nitrate, an excellent nitrogen source for plant growth. The kinds of sites in which nitrate accumulates in nature have been known for many centuries. These include compost heaps, manure piles, burial mounds, and guano deposits—in short, locales rich in organic matter.The best accumulations are found in geographic areas that are warm and relatively dry because nitrate salts are very soluble in rainwater and, consequently, are readily washed away.

Potassium nitrate (saltpeter) was in use by humans long before biology and chemistry became sciences. Originally, it was simply recognized as some sort of matter that could be easily extracted with water from compost heaps and animal wastes. When the water was removed by evaporation, the residue had the interesting property that it formed an explosive black powder when mixed with charcoal (carbon) and sulfur in these proportions: residue, 75 percent; charcoal, 15 percent; and sulfur, 10 percent. This recipe was first used (as far as we know) in China in about the tenth century for fireworks at celebrations and for signaling at a distance. In the fourteenth century, Europeans discovered that this black powder could be used as gunpowder. Enormous deposits of sodium nitrate (Chile saltpeter) were discovered in Chile during the 1830s. Sodium nitrate is not quite as explosive as potassium nitrate, but it is easy to convert the sodium form to the potassium salt.

This excerpt, which describes the production of gunpowder from saltpeter, shows an early example of microbial technology.*

> All the sites in which saltpetre was found were protected from the sun and rain, and contained large quantities of nitrogenous organic matter. So, when Napoleon wished to have saltpetre made in France for gunpowder, because the blockade had cut off the supply of imported nitrate, nitre-heaps were made in imitation of the natural sites (J. B. Boussingault: *Economie rurale.* Paris: Dechet Jeune, 1844). Heaps of a mixture of earth, manure, and chalk [the latter to supply CO_2] were built inside wooden sheds; they were watered with urine [source of ammonia] and waste water, and either aerated by pipes or turned from time to time. After about two years the crude saltpetre was extracted from the heaps with hot water.

DENITRIFICATION

It was noted earlier that nitrate is an excellent nitrogen source for plant growth, and consequently, microbial production of this form of nitrogen in soils (nitrification) is of agricultural significance. However, if anaerobic conditions become estab-

*From Meiklejohn, J., 1953, The nitrifying bacteria: A review. *Journal of Soil Science*, 4:59.

lished in soil, through compaction of the soil or by waterlogging, the availability of nitrate is diminished because of the activities of denitrifying bacteria. A number of species of heterotrophic bacteria have the ability to use nitrate as a substitute for O_2 in energy metabolism, through a variation of bioenergetics known as *anaerobic respiration.* When nitrate is used in this way, it is converted to gaseous forms of nitrogen, principally N_2 (some nitrous oxide is also released to the atmosphere). Soils frequently contain as many as 1,000,000 denitrifying bacteria per gram, and in waterlogged conditions as much as 15 percent of the inorganic nitrogen present may be lost by conversion of nitrate to N_2. The practice of tilling soil inhibits denitrification by increasing exposure of the microbial flora to O_2 in the air. Pockets of anaerobiosis develop within large soil clumps, whereas smaller particles permit better diffusion of O_2 and are consequently more aerobic. When both nitrate and O_2 are available, denitrifying bacteria preferentially use the oxygen gas and ignore the nitrate.

BIOLOGICAL N_2 FIXATION

At the end of the nineteenth century,Sir William Crookes, a famous British chemist, painted a doomsday scenario of the future world in which food production would collapse because of the lack of nitrogen fertilizer for growing crop plants. In Crookes' time the main source of nitrogen fertilizer was the Chilean nitrate deposits, and he predicted that these would soon be exhausted in the efforts to feed multiplying populations in the industrial countries. Fortunately, ammonia is as good as nitrate in providing nitrogen for plant growth, and the natural process of bacterial nitrogen (N_2) fixation represents the possibility of a virtually inexhaustible supply of ammonia and useful organic nitrogen compounds. *Nitrogen fixation* is defined as the conversion of gaseous (atmospheric) N_2 to ammonia and organic nitrogen and is the "final" phase of the Earth's nitrogen cycle we will consider. Obviously, N_2 fixation has been playing an important role in the recirculation of nitrogen atoms on Earth for millions of years. It is only in the past several decades that we have realized the possibility of exploiting bacterial N_2 fixation in a major way for improving agricultural efficiency.

The fixation of N_2 by bacteria can be summarized by the simple equation:

$$N_2 + 6H \text{ atoms} + \text{energy (ATP)} \xrightarrow{\text{nitrogenase}} 2NH_3$$

Bacteria catalyze this overall reaction at ordinary temperatures and gas pressures through operation of a remarkable enzyme system called *nitrogenase*. Nitrogenase consists of two enzyme proteins that act in concert; both of them contain metals that are essential for their enzymatic activities. One of the proteins depends on the presence of attached iron atoms and the other on iron and molybdenum atoms. In addition to nitrogenase, the conversion of N_2 to ammonia requires provision of hydrogen atoms from metabolites, and a source of energy, namely ATP. Molecular nitrogen (N_2) is relatively inert because its two constituent N atoms are held together by strong chemical bonds, in fact, *three* chemical bonds (N≡N). The nitrogenase proteins are specially designed to accomplish the separation of the two N atoms and their conversion to ammonia under mild conditions.

Ammonia (NH_3) is used for the synthesis of amino acids (and thus, proteins) and of all other nitrogenous constituents of cells in much the same fashion throughout the living world. The only unique aspect of N_2-fixers is the transformation of N_2 to NH_3. In 1913, Fritz Haber* (1868–1934) and Carl Bosch (1874–

*Fritz Haber was an eminent German physical chemist whose most outstanding accomplishment was development of a commercially feasible artificial N_2 fixation process (1913). During World War I, he devoted his efforts to war-related projects; by that time, the Haber–Bosch process was being used to supply Germany with nitrogen compounds needed for fertilizers and explosives. He was awarded the Nobel Prize in Chemistry in 1919. Later, as director of the Kaiser Wilhelm Institute for Physical Chemistry and Electrochemistry, he de-

1940), two German chemists, developed a chemical method for producing artificial nitrogen fertilizer in the form of ammonia. Currently, most of the nitrogen fertilizer deliberately added to farm soils is applied as ammonia, made in factories by the Haber–Bosch process. This process is represented as follows:

$$N_2 + 3H_2 \xrightarrow[\text{catalysts}]{\text{iron or nickel}} 2NH_3$$

Special inorganic catalysts containing iron or nickel are required to produce the resultant ammonia. The gaseous H_2 is made from natural gas which consists largely of methane (treatment of methane with steam yields H_2). For N_2 and H_2 molecules to react on the surfaces of the catalysts, the gases must be heated to about 900°F (480°C) and kept under pressures about 1000 times that of normal atmospheric pressure. In other words, artificial N_2 fixation as currently practiced is a high-technology process that requires much energy and engineering. If we could understand the intimate details of how N_2 fixers accomplish the same chemical conversion at ambient temperature and pressure, it is likely that a more economical artificial process could be designed. This expectation is the basis for intensive worldwide efforts during the 1980s to unravel the mechanisms used by biological N_2 fixers.

Thus far, only bacteria have been mentioned in connection with N_2 fixation. Indeed, the only organisms known to have this capacity are bacteria. The occurrence of N_2 fixation ability in different bacterial genera with a variety of lifestyles seems haphazard. One interpretation of this fact suggests that the genes controlling formation of the N_2 fixation system are probably transferred readily in nature between different kinds of bacteria. This is no small feat when we consider that a large as-

---

veloped a world-famous scientific research center. When the Nazis came to power and demanded dismissal of Jewish workers in universities and research institutes, Haber (himself Jewish) resigned his post. He was offered a position as head of the physical chemistry section at the Daniel Sieff Research Institute in Israel and was on his way to the opening ceremonies for the Institute when he died in Switzerland.

sembly of genes, 17 in all, is needed to provide the genetic blueprints for biosynthesis of the active nitrogenase machinery. This gene transfer has actually been accomplished in the laboratory between certain bacterial species. Is it possible that the nitrogenase "gene family" could be introduced into the genetic apparatus of wheat, corn, or other important crop plants? If so, and if the gene family would actually function as it does in a typical N_2-fixing bacterium, the genetically engineered plant could presumably be grown without adding artificial N-containing fertilizers. This possibility is the stimulus for much current research in biotechnology. There may, however, still be unknown biochemical obstacles that will prevent the expression of the nitrogenase genes in a eucaryotic plant cell environment.

ECOLOGY OF N_2 FIXERS

Many genera of bacteria include species that are freeliving N_2 fixers; that is, organisms that live and grow as unicellular forms in soil, natural waters, and other habitats. Many of these are heterotrophs, anaerobes as well as aerobes. Interestingly, of the 60 known species of anaerobic photosynthetic bacteria (see Chapter 13), all but one fix N_2 using light as the source of energy. Nitrogen fixation is also widespread among genera of aerobic photosynthetic bacteria (the cyanobacteria). Some scientists speculate that bacterial N_2 fixation is a very ancient biochemical system that became operative before the Earth's atmosphere contained any O_2 and while the Earth was still populated only by anaerobic procaryotes (before the evolution of plants). There is much to commend this interesting idea, particularly the fact that activity of nitrogenase is strongly inhibited by O_2 when it is present at the concentration now found in the atmosphere (about 21 percent; N_2 accounts for approximately 78 percent). Aerobes that fix N_2 have special biochemical devices to diminish the O_2 pressure in the immediate vicinity of nitrogenase, and this is particularly evident in the kind of N_2-fixing system described below.

Figure 14 The effect of N_2-fixing *Rhizobium* bacteria on growth of alfalfa plants. The plants are growing in nitrogen-poor soil. *Rhizobium* cells were added to the soil in the right-hand pot.

Fortunately, nature has devised ways to obtain "semianaerobic" conditions in the presence of what might appear to be aerobic circumstances. The most interesting, and perhaps most important, arrangement is found in *symbiotic* N_2-fixing systems. These are typified by legume plants such as soybeans, clover, and alfalfa. Root hairs of legumes become naturally infected with N_2-fixing bacteria of the genus *Rhizobium*. This leads to the formation of small nodules on the roots in which the *Rhizobium* cells grow abundantly. When a nodule is crushed and its fluid observed under the microscope, millions of *Rhizobium* cells are observed. Within the nodules of a growing plant, the O_2 pressure is kept relatively low by biochemical systems of the plant cells. This ensures optimal conditions for N_2 fixation by *Rhizobium*. Thus, the bacteria–plant relationship is symbiotic. The bacteria in their nodule environment are provided with carbon sources and other metabolites by the host plant, and they fix enough N_2 to provide themselves and the plant with the ammonia needed for growth. The benefit to the plant is easily demonstrated by simple experiments (Figure 14).

There are other kinds of symbiotic N_2-fixing systems, for example, a type in which nodules are formed on leaves rather than on roots. In all instances, however, the fundamental biochemical aspects of the process are essentially the same. Those readers interested in a recent overview of technical details should consult works by J. R. Postgate, a leading researcher in the field.*

*See, in particular, Postgate, J. R., 1982, Biological nitrogen fixation: fundamentals. *Phil. Trans. Roy. Soc. London (ser. B)*, 296:375.

11

Bacteria Spin the Sulfur Cycle

In the normal metabolism of living cells, inorganic forms of sulfur are converted to organic forms. Two of the twenty amino acid "building blocks" of proteins contain sulfur atoms, and sulfur is also a constituent of several vitamins. When plants and animals die, organic sulfur compounds are decomposed by bacteria with the release of hydrogen sulfide (H_2S), an inorganic form of sulfur with an obnoxious smell. Sulfur occurs on the Earth in several other inorganic forms, all of which are constantly being interconverted on a massive scale. Bacteria are active agents in most of these processes, and it is not an exaggeration to say that bacteria "spin" the sulfur cycle.

We will consider a simplified version of the sulfur cycle, paying particular attention to the three inorganic forms: sulfide, elemental sulfur (S), and sulfate (SO_4).* In the skeleton cycle

*Sulfur dioxide (SO_2), an inorganic form of sulfur not considered here, is produced during burning of fossil fuels, oil processing, and smelting of iron ores. The acid rain problem is caused by sulfurous acid that is generated when SO_2 dissolves in rain and other waters.

shown below, sulfide is represented as the compound hydrogen sulfide (H_2S) and sulfate is in the form of sulfuric acid (H_2SO_4).

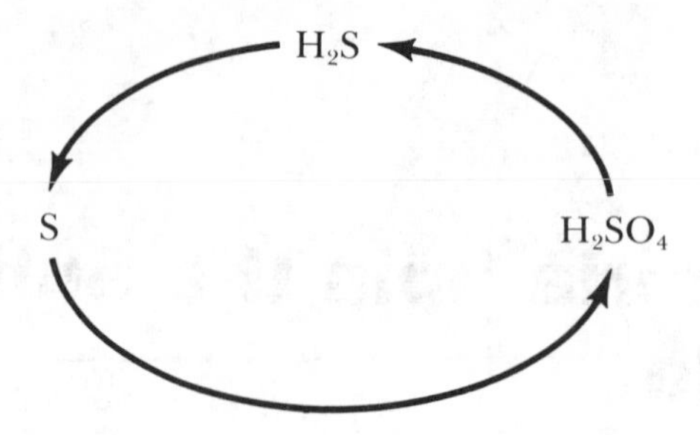

Note that the three sulfur forms shown differ with respect to how many H and O atoms are associated with the S atom. In sulfate, each S is bonded to four O atoms, and this is the case regardless of the type of sulfate: for example, sulfuric acid (H_2SO_4), calcium sulfate or gypsum ($CaSO_4$), and magnesium sulfate or Epsom salt ($MgSO_4$). All natural waters contain sulfates, and on a global scale, huge amounts of sulfate are converted to H_2S by anaerobic bacteria in soil, mud, lake beds, sewage, and other aqueous environments. One prominent species of bacteria almost always at work in this part of the sulfur cycle is *Desulfovibrio vulgaris*. This bacterium has numerous "cousins" in genera with similar-sounding names: for example, *Desulfobulbus* and *Desulfobacter*. These organisms convert sulfate to hydrogen sulfide to obtain energy (ATP). In effect, they use sulfate, a highly oxygenated form of sulfur, as a substitute for gaseous O_2. In fact, they are strict anaerobes and use sulfate only in natural situations in which O_2 is absent (or present in very low amounts). The energy-yielding systems of the "Desulfo's" can be considered as a variation of anaerobic respiration, analogous to the anaerobic use of nitrate (as an O_2 substitute) by denitrifying bacteria.

The accumulation of H_2S due to the activities of the "Desulfo's" is of considerable economic significance because sulfide gradually causes corrosion of metals, for example, of buried pipelines. In addition, H_2S in moderate concentrations poisons the enzymes of aerobic respiration in animals and other aerobes; in fact, it is as poisonous as cyanide.

In many natural situations, however, H_2S produced from sulfate does not accumulate to significant levels because it is removed by the metabolism of other kinds of bacteria. As indicated in the diagrammatic cycle, the H_2S is converted to elemental sulfur (S) and eventually to sulfate. Depending on the circumstances, two kinds of bacteria are concerned with reformation of sulfate. Under anaerobic conditions, the transformation of H_2S to sulfur and sulfate is accomplished by photosynthetic bacteria, pigmented organisms (containing chlorophyll) that use light energy to replenish their ATP supplies (discussed in Chapter 13). These autotrophic bacteria avidly use the H atoms of H_2S for the production of sugars and other cellular compounds from CO_2. A representative ecological situation in which the anaerobic part of the sulfur cycle predominates has been described by John Postgate as follows.* (Note that in this quotation, "reduce" means to remove oxygen atoms from sulfate; "sulfate-reducing bacteria" can remove oxygen from sulfate and thereby produce sulfide.)

> In Libya there are a number of lakes (near the hamlet of El Agheila) where warm artesian water, rich in calcium sulphate and containing hydrogen sulphide, comes to the surface through springs. One of these, called Ain-ez-Zauia, is about the size of a swimming pool and is slightly warm (30 degrees centigrade). It is saturated with calcium sulphate and contains about 2.5 percent sodium chloride—a reasonable approximation to a warm, drying-up sea, if a little weak in salt. Under the Libyan sun, this lake produces about 100 tons of crude sulphur a year, formed as a fine, yellow-grey mud which is, in fact, harvested by the local Bedouin. (They export some to Egypt—or did when I was there in 1950—and use it as medicine themselves.) The way in which the sulphur is formed is this: sulphate-reducing bacteria reduce

---

*From Postgate, J., 1986, *Microbes and Man*, 2nd ed., Penguin Books, Baltimore.

> the dissolved sulphate to sulphide at the expense of organic matter formed by coloured sulphur bacteria, which in their turn have made the organic matter from carbon dioxide using sunlight and sulphide, some of it from the spring waters, some formed by the sulphate reducers. Thus we have sulphur formed from sulphate by two interdependent types of bacteria, the whole process being propelled by solar energy. The bed of the lake consists of a red, gelatinous mud made up almost entirely of coloured sulphur bacteria; the bulk of the lake is a colloidal suspension of sulphur rich in sulphate-reducing bacteria; the whole system smells strongly of hydrogen sulphide.

Other kinds of bacteria involved in recycling sulfur are aerobic, with the genus *Thiobacillus* being particularly noteworthy. Bacteria of this genus are found in all soils and in the acidic water that drains from various kinds of mining operations. *Thiobacillus ferrooxidans*, an important representative, was first isolated from flood waters draining from an abandoned coal mine in West Virginia. This bacterium adsorbs (attaches itself) tenaciously to the mineral pyrite (FeS_2) and converts it, in the course of energy metabolism, to a mixture of iron sulfate and sulfuric acid. Pyrite usually occurs in association with more valuable minerals such as copper and uranium, and the action of *T. ferrooxidans* greatly increases the solubility of these metals. This facilitates mining, and Canadian companies are now using *T. ferrooxidans* and similar bacteria on a large scale for bacterial leaching of minerals.

The production of sulfuric acid from sulfide by thiobacilli can, however, have deleterious consequences. Concrete that contains appreciable levels of sulfide is subject to deterioration by such bacteria. There have been instances of total collapse of concrete cooling towers caused by *T. thiooxidans* (originally named *T. concretivorus*). Similar problems have occurred in tropical climates where stone buildings were erected on sites rich in sulfide. Absorption of hydrogen sulfide by porous stone and its subsequent conversion to sulfuric acid by thiobacilli is believed to be causing gradual destruction of the temple ruins at Angkor Wat in Cambodia.

12

An Amazing Diversity of Lifestyles

Ecology is the branch of biology that deals with the relationships of organisms to one another and to their surroundings. In comparison with other kinds of living creatures, microbes are extraordinary in respect to the great diversity of ecological niches in which different species can grow. In other words, in the microbial world a particularly wide range of chemical and physical conditions can be tolerated and exploited. Although the basic biochemical processes of all microbes and other organisms are much the same, some species of microbes have alternative patterns superimposed on the fundamental metabolic framework, and these permit growth in extreme environments (see Figure 15). Extremes can occur in such physical conditions as temperature, high concentrations of salts and sugars, relative acidity, and absence of oxygen (the latter of which was discussed in Chapter 3). A recent text on environmental microbiology (see Suggested Readings) notes that extreme environments present a fascinating challenge and provide numerous ecological and biochemical insights into the extraordinary conditions under which life can flourish.

Figure 15 *Translation: "Oxygen content is 21 percent, only a trace of CO_2, and no methane–There can't be any life forms on this planet!"

TEMPERATURE EXTREMES

Some Like It Hot The upper temperature limit for growth of multicellular animals and plants is about 50°C (122°F). Above that temperature, the only forms that can grow are certain species of microbes. A number of bacterial species can even grow in boiling water at 100°C (212°F) and some display an optimum growth temperature of 85°C (185°F). Microbes with such capabilities are called *thermophiles,* and they are found in naturally hot environments such as compost heaps and hot springs (and sometimes in artificial environments such as hot water heaters).

The optimum temperature for growth of any kind of cell is related to the temperature stability of important classes of macromolecules (especially proteins and nucleic acids) and to the effects of temperature change on the hundreds of enzyme reactions necessary for synthesis of new cell material. Most types of cells or organisms grow best in the temperature range 25 to 45°C (77 to 113°F) (so-called *mesophiles*), and when the temperature is increased above 50°C (122°F), their enzyme proteins are adversely affected. In fact, they may coagulate, as egg white protein does when heated. Thermophiles are distinctive in that their enzymes are stable at high temperatures; indeed, they work better in hotter environments. This property has attracted considerable interest in the use of thermophiles and thermophilic enzymes in industrial biotechnology. Chemical processes catalyzed by mcirobes, such as fermentations, frequently release energy in the form of heat, and expensive cooling procedures are needed when using mesophilic microbes to maintain temperature in an acceptable range. With thermophiles this is unnecessary, and at high growth temperatures problems with contamination by unwanted mesophilic microbes are minimized. One thermophilic process now under development is the fermentative production of alcohol from cellulose by *Clostridium thermocellum.*

The unusual temperature stability of many thermophilic bacteria is reflected in many of their generic and species names: *Thermus thermophilus, Bacillus thermoproteolyticus, Thermoanaerobacter ethanolicus,* and so on. The upper temperature limit for thermophilic bacteria, as far as is known at present, is about 110°C (230°F). (At atmospheric pressure, water boils at 100°C;

to achieve a liquid temperature of 110°C, the medium must be under greater pressure.)

Others Prefer It Cold The enzyme activities, and thus the growth, of most kinds of cells slow down markedly as temperature decreases. *Psychrophilic* microbes, however, prefer lower temperatures (the Greek word for "cold" is *psychros*). A true psychrophile is defined as an organism with an optimal growth temperature of 15°C (59°F) or lower, and a minimal temperature for growth of 0°C (32°F) or lower. There are many natural locales that have low temperatures most of the time; for example, the depths of the oceans are at about 1 to 2°C (34 to 36°F) and large areas of the Arctic and Antarctic regions are permanently frozen. Microbes have been found alive and well in such places. Recent studies have revealed extensive growth of psychrophilic oxygenic photosynthetic bacteria on the bottom of shallow ice-covered lakes in Southern Victoria Land, Antarctica. These lakes, which have a temperature close to 0°C, lack outflow streams but receive a limited supply of glacial meltwater that carries nutrients and salts. A prominent organism in these cold ecosystems is appropriately named *Phormidium frigidum*. Despite the low temperature and the relatively low light intensity at the lake bottoms, *P. frigidum* grows abundantly. In order to reach these organisms the intrepid microbiologists who study the ecology of these lakes must scuba dive into the lake via holes bored through 18 feet of ice using a large copper coil with hot antifreeze pumped through its tubing.*

OSMOPHILES

Natural environments that contain high concentrations of salts or other small molecules pose problems for ordinary microbes.

---

*See Young, P., 1981, Thick layers of life blanket lake bottoms in Antarctica valleys. *Smithsonian Magazine*, (November) p. 52.

Within any kind of cell, dissolved salt molecules and small organic molecules are in constant motion and collide with the inside surface of the cell wall, creating an internal pressure. The intensity of the molecular bombardment of the cell wall interior depends on the total number of small molecules dissolved in a unit volume, and this value determines what is referred to as the internal *osmotic pressure.* Similarly, the growth medium has a characteristic osmotic pressure, again dependent on the total concentration of small molecules. Because of certain laws of chemistry and physics and the behavior of cell walls with respect to the movement of water and other molecules, two potential situations may cause cell death:

1. If the medium has a considerably lower osmotic pressure than the cell interior, water molecules will *enter* the cell. The cell will consequently swell and eventually burst, causing death.
2. If the medium has a significantly higher osmotic pressure than the cell interior, water will *leave* the cell. Eventually, the cell will shrivel up and die.

Because of these effects, most microbes can only grow in media that have low concentrations of salts and nutrient molecules. This is the basis of the practice of preserving certain foods by adding lots of salt or sugar. Some microbes, however, have the capacity to grow in solutions that contain very high concentrations of salts or other small molecules. These are known as *osmophiles* or *halophiles* (*halos* is Greek for "salt"). They grow selectively in both artificial and natural fluids that have very high osmotic pressures. Salt is often produced commercially from seawater (which contains about 3.5 percent "table" salt) by evaporation in large basins called "salt pans." As the water evaporates, the salt concentration increases to as high as 25 percent, and the salt then crystallizes out of solution. Salt pans usually develop a pink or red color due to the growth of pigmented osmophilic microbes. Bacteria are even found in unusual natural habitats which contain extremely high salt concentrations (as high as 29 percent), such as the Great Salt Lake in Utah and the Dead Sea in Israel.

The Dead Sea is of special interest because of the long history of observations on this awesome sea, set in a desolate landscape where the summer heat is scorching. The earliest report on absence of life in the Dead Sea is believed to be in a book by Aristotle (384–322 B.C.). He wrote that the sea was so "salty bitter" that fish could not live in it. Gradually, the Dead Sea acquired a mystical and sinister reputation, and was described by Thomas Fuller (see Figure 16) in 1650 as follows:

> This *Salt-sea* was sullen and churlish, differing from all other in the conditions thereof. David speaking of other seas, saith, 'there goe the ships, and there is that Leviathan which thou hast made to play therein': so instancing in the double use of the sea, for ships to saile, and fishes to swim in. But this is serviceable for neither of these intents, no vessels sailing thereon, the clammy water being a reall Remora [something that holds back] to obstruct their passage; and the most sportfull fishes dare not jest with the edged-tools of this Dead-sea; which if unwillingly hurried thereinto by the force of the stream of Jordan, they presently expire. Yea, it would kill that Apocrypha-Dragon, which Daniel is said to have choaked with lumps of pitch, fat, and hair, if he should be so adventurous to drink of the waters thereof; so stiffling and suffocating is the nature of it. In a word, this sea hath but one good quality, namely, that it entertains intercourse with no other seas; which may be imputed to the providence of nature, debarring it from communion with the Ocean, lest otherwise it should infect other waters with its malignity. Nor doeth any healthfull thing grow thereon, save onely this wholesome counsell, which may be collected from this pestiferous lake, for men to beware how they provoke divine justice, by their lustfull and unnaturall enormities.

Much later, an expedition in 1861–1863 reported that despite all efforts, "no living creatures were found in the waters of the Dead Sea proper." Finally, in 1940 Benjamin Volcani demonstrated conclusively that the Dead Sea contains many

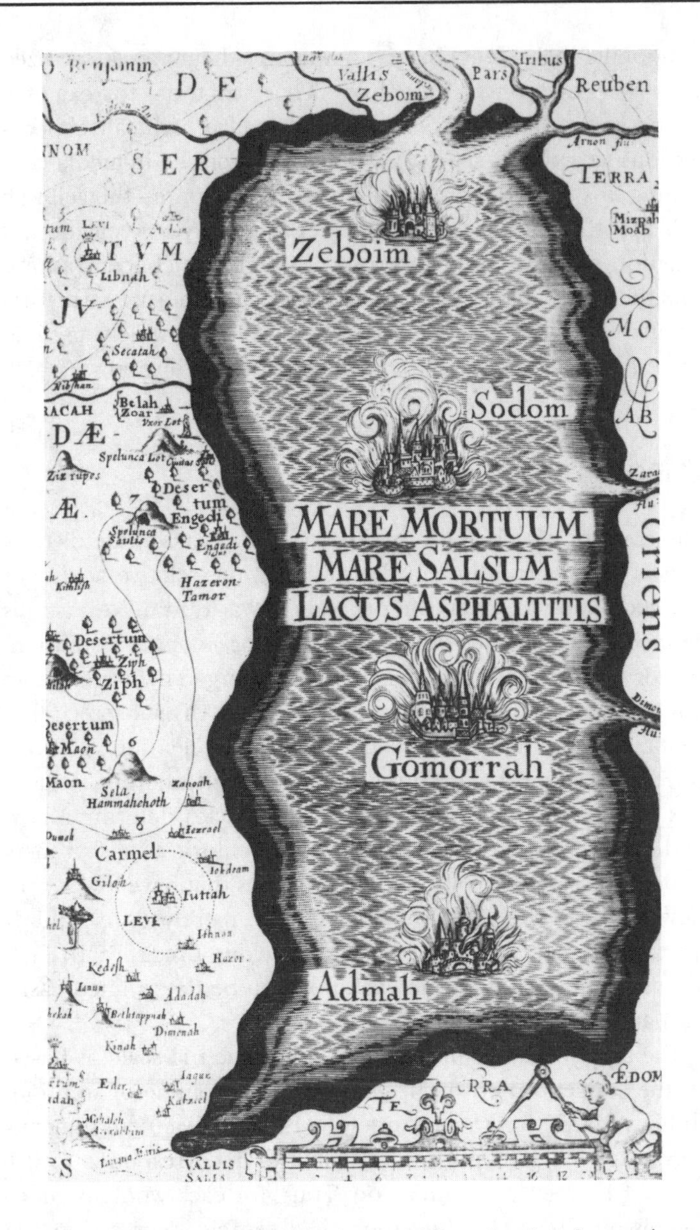

Figure 16 An ancient map of the Dead Sea, showing destruction of the "sinful cities of the Plain."

osmophilic (halophilic) bacteria. Some of them are categorized as *extreme*, or *obligate*, halophiles. These bacteria, for example of the genus *Halobacterium*, actually require high salt concentrations to grow. If *Halobacterium* cells are placed in media containing less than 10 percent salt, they disintegrate—the cell wall falls apart! A number of enzymes in *Halobacterium* cells do not function properly unless exposed to high salt concentrations, and the cell walls of these organisms have unusual structural features. A species of *Halobacterium* which is widely used in current research is *H. volcanii* (named after Volcani).

ACIDOPHILES

The concept of acidity is well known: for example, the distress caused by "overacidity" in the stomach, the relative acidity of shampoo and soap, and the problem of "acid rain." An acid is defined as a substance that liberates *hydrogen ions* in a solution. A hydrogen ion is simply an electrically charged hydrogen atom (the charge is positive), represented as H^+. Any substance that contains H atoms is potentially capable of releasing H^+ ions. Different chemical substances have different inherent tendencies to release H^+ ions, and these tendencies determine how acidic they are. Thus, concentrated hydrochloric acid is strongly acidic, and very corrosive. Acidic solutions that have lower concentrations of H^+ ions, such as vinegar, merely taste sour. It is thus necessary to measure relative acidity precisely. In fact, the degree of acidity of any solution can be determined easily with appropriate devices.

The acidity measurement scale, called the pH scale, is based on the concentration of H^+ ions in a unit volume of solution, and has values that range from 0 to 14. A pH value of 0 means 1 gram of H^+ ions per liter, a value of 1 corresponds to 0.1 gram of H^+ per liter, and so on. Thus, for each whole number increase in the pH value, there is a tenfold decrease in the concentration of H^+ ions. Even in the purest water, there are

Table 7 Relative Acidity of Various Liquids

	pH (approx.)	Solution or Environment
↑	1.0	Automobile battery acid; human gastric juice
	2.3	Lemon juice
Increasing	3.0	Vinegar
acidity	4.7	Rain in most of the eastern United States
	5.0	Sour milk
	6.3	Adirondack Lakes, New York, in 1930
....	7.0	Distilled water
	7.4	Human blood plasma
	8.4	Seawater; baking soda solution
Increasing	10.0	Great Salt Lake, Utah
alkalinity	12.0	Ammonia water
↓	13.0	Lye (caustic soda)

some H^+ ions. . .for about every 550 million water (H_2O) molecules, there will be one that releases an H^+ ion. This corresponds to a pH value of 7 for water, and 7 is considered to be the neutral point of the scale. Solutions that have pH values less than 7 are said to be acidic; values greater than 7 are labeled as alkaline or basic. It is instructive to consider some characteristic pH values (Table 7).

Most microbes grow best in the pH range of 6 to 8, close to the neutral zone, and cannot grow in media with pH values as low as 3 to 4. This fact is the basis for pickling food as a means of preserving it. Various vegetables and other foods can be preserved by adding vinegar or other acidic solutions that are edible and nontoxic. However, certain microbial species called *acidophiles* are well adapted to life in acidic environments. The water that drains coal, copper, and zinc mining operations usually has a very low pH, and acidophilic bacteria have been found growing at a pH of about 1.5 in this unlikely niche (for example, *Thiobacillus ferrooxidans*, an autotrophic acidophile). There is evidence that acidophiles maintain an internal pH between 6 and 7, even when the growth medium is much more acidic. This is achieved in such organisms by constant extrusion (pumping out) of H^+ ions across the cell envelope.

During the past decade, a number of thermoacidophilic bacterial species were isolated for the first time. These unusual organisms, typified by *Thermoplasma acidophilum*, are remarkable because of a simultaneous requirement for *two* kinds of extreme conditions: elevated temperature and low pH. *Thermoplasma acidophilum* is a heterotroph that lives in self-heating coal refuse piles.

As might be expected, certain bacteria prefer alkaline pH conditions. Alkalophilic photosynthetic bacteria of the genus *Ectothiorhodospira* inhabit alkaline soda lakes in East Africa. These lakes, which have a pH of about 11, are commercially exploited for the mineral sodium sesquicarbonate (trona).

MAGNETOTACTIC BACTERIA

The magnetic poles of the Earth correspond closely to the geographic north and south poles. Thus, any device that automatically points to the north magnetic pole would obviously be very useful for direction finding, especially on the high seas. In the twelfth century, Italian ship pilots began to check the direction of north using a magnetized iron needle floating in a bowl of water. The needle was magnetized by rubbing it with a piece of magnetite, a natural iron oxide mineral then known as lodestone. During the next century, this practice led to the development of the mariner's compass. In 1975, the surprising discovery was made that some types of bacteria behave as if they were swimming bar magnets. In other words, they swim as if they contain internal compasses that direct the cell toward one of the Earth's magnetic poles.

Amazingly, *magnetotactic* bacteria that are found in the Northern Hemisphere swim toward the Earth's pole that attracts the north-seeking end of a compass needle. Their counterparts in the Southern Hemisphere (found in Tasmania and New Zea-

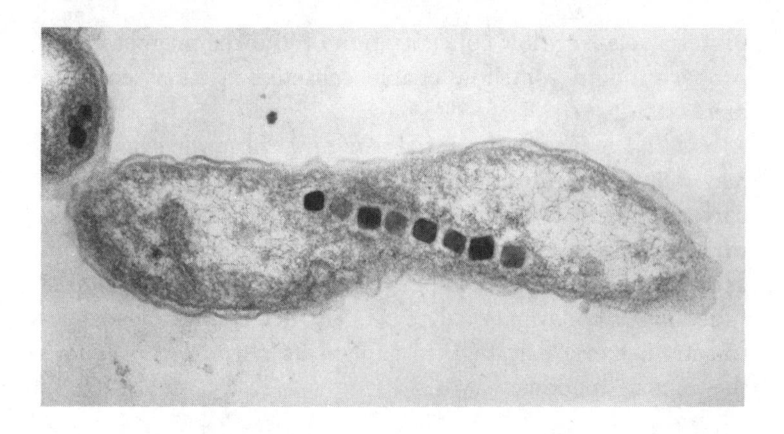

Figure 17 Electron micrograph of the magnetotactic bacterium *Aquaspirillum magnetotacticum*. The cell contains magnetite particles lined up in a chain-like array.

land) swim toward the south-seeking direction. Magnetotactic bacteria behave as bar magnets because each cell contains either one or two chains of magnetite particles (Figure 17). The magnetic particles are present only if the bacteria grow in an environment that contains sufficient soluble iron; with low levels of iron, the bacteria may still grow but they do not make magnetite particles and are not magnetotactic.

As to the function of the magnetotactic behavior, there are several possibilities. Formation of the particles could simply be a way of getting rid of excess iron. Another reason may be that the particles assist the cell in disposing of hydrogen peroxide, which is a toxic product of the metabolism of O_2 gas and is effectively destroyed by magnetite. An appealing alternative is that magnetotactic behavior represents a mechanism that the bacteria use to find a suitable ecological niche, particularly with respect to the environmental concentration of O_2 gas. These organisms are typically found in or near muddy sediments, such as in bogs or marshes. Between 100 and 1000 magnetotactic bacteria are found per milliliter of such materials. Furthermore, they are either anaerobic or "microaerophilic"; that is, they prefer a relatively low concentration of O_2. The magnetotactic property might somehow enable cells to find such environments.

It would seem that bacterial wonders will never cease! One may well ask: Is magnetotactic movement unique to the procaryotic kingdom? Probably not. Magnetite has been found in various animals including honeybees, butterflies, homing pigeons, and dolphins. It is posible that the magnetic material functions as a navigational guide. Further studies on magnetotactic bacteria may help to explain its purpose throughout the animal kingdom.

MICROBES IN DEEP SEA VENTS

The symbiotic N_2 fixation system of legumes is one example of microbes living in close association with higher organisms. Depending on the nature of the relationship between microbe and host, such associations have different consequences. The relationship may be neutral in character, that is, with no apparent advantage or disadvantage to the microbe or the biological host. Alternatively, the association may be of mutual benefit, or the microbe may be parasitic and thereby cause deterioration of the host. Unanticipated discoveries made during the past decade have revealed the existence of fascinating microbe–animal associations in an unusual marine environment.

In 1977 geologists and geochemists embarked on a program of deepwater exploration to study how new crust is formed along ridges on the ocean floor. The U.S. Navy submersible *Alvin* made dives to depths of about 2500 meters along the Galapagos Rift at the equator (at about longitude 86°W). The *Alvin* carries one pilot and two observers and is equipped with one mechanical and one hydraulic arm for placing and operating instruments and collecting samples.* Along the rift, the

---

*The Alvin was also used for finding and exploring the remains of the *Titanic*

scientists discovered hydrothermal vents—in essence, underwater volcanoes. The vents spew out very hot water (hydrothermal fluid) and gases, which include hydrogen sulfide (H_2S), methane (CH_4), carbon monoxide (CO), and hydrogen (H_2). As the hydrothermal fluid is emitted, it mixes quickly with cold seawater that contains O_2 and a characteristic assortment of inorganic salts.

Some of the vents found by the *Alvin* explorers emit jets of superheated black water (as hot as 350°C or 660°F) from chimneys formed of metallic sulfide minerals, so-called black smokers. What surprised the geologists and geochemists the most was the observation of dense animal communities growing near the hot spring outlets, in total darkness. All animal life on and in the surface layers of the Earth is ultimately dependent on plant organic matter and thus on photosynthesis. How, then, were these unique communities able to survive? They are not just surviving, but are thriving in high biomass density: giant clams, huge tube worms, mussels, crabs, undescribed "vent fish," and many other forms. Remarkably, a number of them proved to be genera or species not previously known. (For example, *Alvinella pompejana* ("Pompeii worm") and *Alvinocaris lusca* (vent shrimp), both named after the *Alvin*.) At least two kinds of "living fossils" (varieties thought to be extinct) were also found.

Before long it became apparent that the food chain supporting the abundant animal life along the vents must be based on primary production of organic matter by bacteria. Although much remains to be learned about the bacterial ecology of the hydrothermal vent localities, some fundamental facts have been established. The main bacterial agents in this undersea world must be autotrophs that grow on CO_2 and obtain energy by respiration of H_2S or H_2, and aerobic species that can grow on methane as the sole source of carbon and energy. Thus far,

---

on the bottom of the North Atlantic Ocean; there is a detailed cutaway drawing of the submersible in the December 1986 issue of National Geographic, pp. 706–707.

most studies have concentrated on the H_2S-dependent species: autotrophic "sulfur bacteria." These, and no doubt other kinds of bacteria, are found in the vicinity of the vents. Many strains of the sulfur bacteria have been isolated and show optimum growth temperatures of 25 to 32°C (77 to 90°F).

The giant clams and tube worms account for most of the animal biomass at vent sites studied up to now, and they grow in areas that appear to be devoid of sufficient particulate or dissolved food materials to support such extensive growth. This suggested that their food source must be "endogenous" (originating from within). It was soon confirmed that tissues of these animals contain endosymbiotic bacteria: bacteria that provide food for the host organism live inside the animal cells! It seems that CO_2, H_2S, and O_2 are absorbed by the animals and transported to the endosymbiotic bacteria which, in turn, grow and furnish organic compounds to the animal cells. Another unusual aspect of this ecosystem is that H_2S is usually toxic to aerobic organisms because it poisons catalysts of the energy-yielding respiratory system. This means that the giant clams and tube worms must have some clever adaptive modifications that permit transport of H_2S to the tissue cells without interference to bioenergetic metabolism. (The bacteria also have to be "immune" to sulfide poisoning.)

Many accounts of the ecosystems of the deep sea vents stress the idea that these systems are unique in depending on geothermal (terrestrial) rather than solar energy; in other words, they are sometimes said to be totally independent of photosynthesis. Such statements are inaccurate in that, as far as we know at present, O_2 gas is required for growth of most of the bacterial species that support the hydrothermal vent food chains. The O_2 is, of course, generated by oxygenic photosynthesis that occurs elsewhere (in surface waters and on land).

SOCIAL LIFE STYLES

Chemical "communication" between different species of microbes is an inherent feature of the carbon, nitrogen, and sulfur cycles. This statement refers to the fact that substances excreted into the environment by one organism can be used as nutrients by other species. It was noted in Chapter 9 that molecular hydrogen (H_2) links several different species in natural microbial ecosystems which are responsible for methane production. For example, in the rumen, the H_2 evolved from bacterial fermentation processes is so avidly used by methanogens that its concentration remains very low in the rumen gas atmosphere.

In other instances in the microbial world, "interspecies chemistry" takes the form of an intimate physical association of two markedly different kinds of cells. One example is a commonly occurring *consortium* between a bacterium we can designate as "A," that converts sulfate to H_2S in its energy metabolism scheme, and another bacterium, "B," that needs H_2S as a source of H atoms. When H_2S is used for the latter purpose, the sulfur is converted to sulfate. Thus:

- "A" converts sulfate to H_2S
- "B" converts H_2S to sulfate

Under the microscope, the sulfate-requiring cell is observed to have a dozen or more of smaller H_2S-requiring bacteria stuck to its surface (see Figure 18). The two species travel together, one piggy-backed on the other, and they feed each other continuously. That is, consortium member "A" converts sulfate to H_2S, and the H_2S is used by "B" in a process that results in its reconversion to sulfate (needed by "A"). Life in a consortium is very economical, and also greatly reduces dependence on environmental supplies of crucial nutrients. In the example given, the sulfur compounds do not even enter the external environment.

A different kind of interdependence among assorted bacterial species is seen in ecological niches where growth occurs in the form of aggregates of cells on solid surfaces. For example, mixed populations of bacteria grow together on the surfaces of teeth, forming the dental plaque. The plaque consists of bacteria that

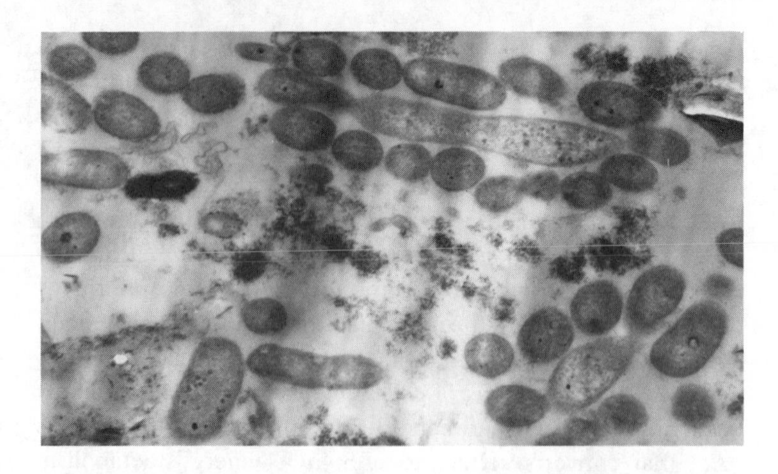

Figure 18 Electron micrograph of "*Chlorochromatium aggregatum,*" a consortium of two different kinds of bacteria that cross-feed each other. The large bacterium in the center converts sulfate to H_2S, whereas the smaller bacterial cells surrounding it metabolize H_2S to sulfate.

are embedded in an amorphous matrix that the bacteria themselves secrete and share. The matrix promotes adherence of the bacteria to the teeth, and is therefore important for their survival. Similar phenomena are encountered in the microbial communities that live on the surfaces of rocks, vegetation, and soil particles. Aggregation presumably enhances the ability of microbes to remain on the surface they are colonizing.

One of the most interesting kinds of "social" interactions of microbes is observed in the growth cycle of individual species of single-celled myxobacteria. These are soil bacteria that live in large populations called swarms. Myxobacteria are natural predators that feed on other microbes, including bacteria. It appears that in their natural habitats, the myxobacteria feed best only when they are in a swarm. In order to digest other microbes, the myxobacteria secrete a battery of enzymes that are capable of breaking down the components of foreign cells. But an individual myxobacterium cannot secrete sufficient digestive enzymes to ensure an adequate supply of nutrients.

Thus, they live only in large populations in which individual cells mutually benefit from sharing the digestive enzymes secreted by other myxobacteria in the swarm. In this way. myxobacteria swarms can rapidly decompose other bacteria in their midst, growing on the products of their victims. Communal feeding by myxobacteria has been likened to the habits of wolves, that can hunt more efficiently as a pack.

When the source of nutrients in a myxobacteria swarm is exhausted, the cell population begins a developmental stage that is unique in the annals of procaryotic microbiology. The individual cells (of a single species) move by gliding in coordinated fashion to "aggregation centers," where they form a complex structure called a fruiting body. Figure 19 is a photograph of a fruiting body of the myxobacterium *Stigmatella aurantiaca*. Within the fruiting body, the myxobacteria eventually form spores. Later, when the fruiting bodies are dispersed to new locations (perhaps by insects) where nutrients are available, the spores germinate and the myxobacteria emerge as a tiny swarm, ready to begin the feeding stage of the life cycle again. The processes of fruiting body formation and sporulation depend on sophisticated systems of cell-to-cell interactions and coordinated movements that are still not understood. At least part of the cycle in certain myxobacteria appears to involve secretion of a "pheromone" (a hormone-like chemical) that promotes aggregation and consequent construction of the fruiting body. Thus, the myxobacteria provide an example of a truly social procaryote, that is, a bacterium whose very survival depends on the ability to synchronize movements of large numbers of cells in order to fabricate a complex structure that facilitates dispersal of the species.

In the preceding, examples were chosen to illustrate the broad range of interaction patterns that occur among microbes in natural habitats. In sum, these include: simple cross-feeding phenomena in which metabolic products (including vitamins) of one organism are used for nutrition of other organisms living in close proximity; secretion of extracellular material (the ma-

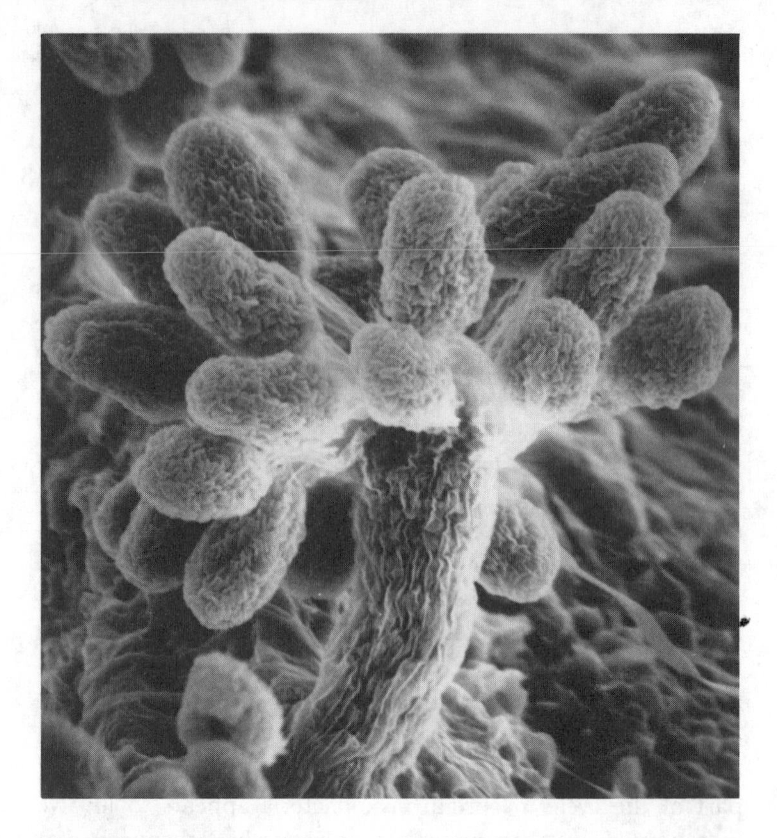

Figure 19 Fruiting body of the myxobacterium *Stigmatella aurantiaca*. Note the multiple "sporangioles" that house the spores, situated on top of a stalk. This fruiting body was found on a decaying tree branch on the campus of Indiana University, Bloomington.

trix) which enables all of the cells in a population to adhere to a surface being colonized; and the sophisticated chemical signaling that takes place between cells for purposes such as aggregation of myxobacteria. As microbiologists continue to investigate the properties of large bacterial populations they no doubt will uncover many examples of phenomena reminiscent of the social behavior of higher organisms.

FUNGI: COMPLEX MICROBES

Up to now, fundamentals of microbial life have been considered using bacteria as the primary examples. In real life, more complex microbes play major roles, for example, in the chemistry of the environment and as pathogenic (disease-producing) agents. *Fungi* (singular, *fungus*) are particularly significant. They are eucaryotes and can be classified as follows:

Microscopic		Macroscopic (multicellular)
Unicellular	*Multicellular*	Mushrooms, toadstools,
Yeasts	Molds	puffballs, and bracket fungi

Two basic features are common to multicellular fungi: (1) under the microscope, cells are observed to occur in the form of threadlike filaments that often have branches, and (2) the filamentous cell masses produce special reproductive structures that shed spores in great abundance. These features are illustrated by the scanning electron micrograph of the mold *Penicillium* shown in Figure 20. (Hereafter the terms *fungus* and *mold* will be used interchangeably.) Spore-heads of different species and genera usually have characteristic features that help in identifying them. Under appropriate nutritional conditions, each spore can germinate and give rise to new filamentous growth and spore-heads.

Molds have essentially the same growth requirements as bacteria, but they ordinarily grow much more slowly. The molds are also distinguished by their capacity to grow on materials that would seem to offer only small supplies of nutrients, for example, linen, cotton cloth, and tanned leather. Thus, molds are found growing in surprising situations, such as on the brine of pickling vats and on fabrics. At least 200 different kinds of molds have been isolated from mildewed fabrics; they grow in patches that often are brightly colored (spore-heads frequently produce pigments that are green, yellow, or brown).

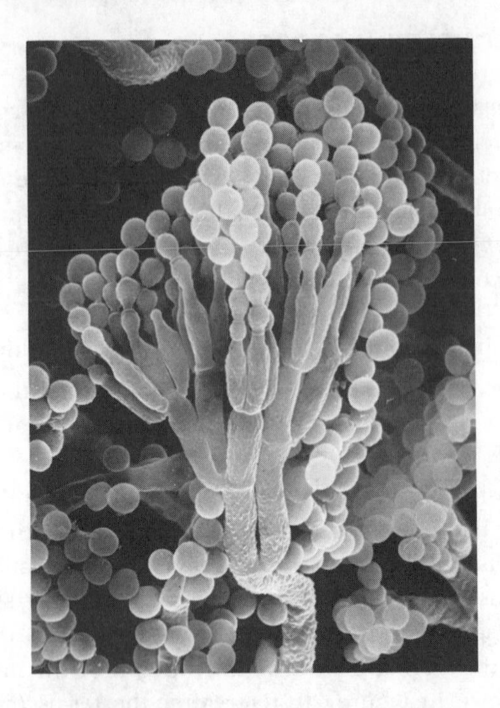

Figure 20 Scanning electron micrograph of *Penicillium*. Note the resemblance of the spore "heads" (bearing chains of spores) to paint brushes. The Latin word for paint brush is *penicillus*. The species here, *P. roquefortii*, is the blue mold used in the manufacture of roquefort cheese.

A strain of the fungus *Trichoderma viride* was isolated from a rotting cotton cartridge belt found in the jungles of New Guinea at the end of World War II. This strain is considered to be potentially valuable for the commercial conversion of cellulose to glucose sugar. The organism has been under investigation by the U.S. Army Natick Research and Development Command, Pollution Abatement Division in Natick, Massachusetts. The Natick scientists have optimistically reported continued improvement of techniques for large-scale production of glucose from various cellulose-containing materials that include urban refuse, agricultural residues, feedlot wastes, newspapers, and "hydropulped'' government documents. The latter seem to be a favorite of the Natick laboratory. Interestingly, Pentagon documents were found to be particularly suitable for production of sugar in a short time. It has been estimated that the cellulose in 1 ton of waste paper can be converted to 0.5 ton of glucose, which can be fermented to 78 gallons of alcohol.

Fungi are also commonly observed growing in colorful patches on barren materials such as bare rocks and house roofs. These patches are growths of *lichens*, which are symbiotic associations of mold with a photosynthetic alga or bacterium. Lichens grow very slowly, but nonetheless can proliferate in extreme environments that offer little organic and other nutrients.

The techniques used for isolating, growing, and studying microscopic fungi are exactly the same as those used for bacteria, and fungi have played important roles in the history of human exploration of the microbial universe. One incident of interest relates to the disease condition known as *convulsive ergotism*. It is caused by eating ergot, which is made by the fungus *Claviceps purpurea* when it grows on rye and other grains. Ergot contains a number of powerful chemicals produced by the fungus, including one substance that is convertible to LSD (lysergic acid diethylamide). The symptoms of ergotism closely match those described in witchcraft trials of children and teenagers in Salem, Massachusetts, in 1692. Court records list the symptoms of ''bewitchment'' as follows: temporary blindness, deafness, or inability to speak; seeing such visions as balls of fire or a multitude of figures in white glittering robes; and the sensation of flying through the air, out of the body. Three girls said they felt as if they were being torn to pieces and as if all their bones were being pulled out of joint. In some severe cases, there was also the sensation of ants crawling under the skin, as well as epileptic-like convulsions. A recent review of the evidence makes it very likely that the victims of the witchcraft accusations were suffering from convulsive ergotism.

13

Bioenergetics: "Energy Currency"

Comparative biochemistry has revealed a number of fundamental similarities in all kinds of cells, no matter what their nutritional idiosyncracies may be. One important similarity concerns the nature of the chemical energy source used for production of cell components by the biosynthetic "machinery." The energy-rich chemical, ATP (adenosine triphosphate), that drives biosynthesis is the same in all cell types and is generated internally from nutrients taken in from the environment. Since different cell types use different nutrients, it follows that there must be several alternative ways in which cells can use nutrient energy to generate ATP. The "energy dynamo" of living cells can be roughly compared to a banking system in which different kinds of valuable materials can be transformed into units of the same kind of "energy currency" ATP.

ATP is a relatively small, but complex molecule containing 47 atoms (carbon, 10 atoms; oxygen, 13; nitrogen, 5; hydrogen 16; and phosphorus, 3). The structure of ATP can be represented in a simplified way to illustrate how the molecule is used

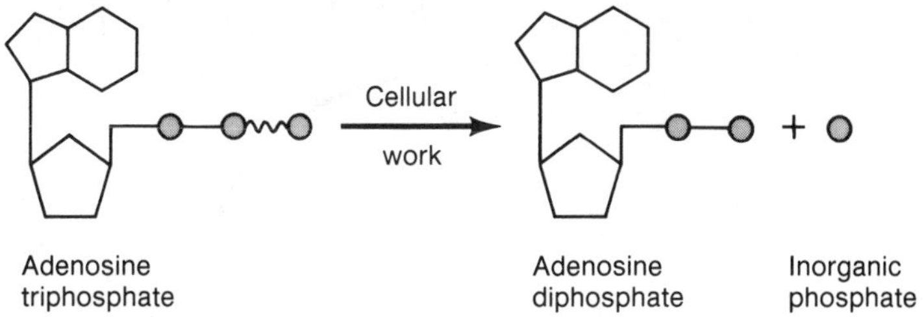

Cellular
work

Adenosine
triphosphate

Adenosine
diphosphate

Inorganic
phosphate

Figure 21 Breakdown of adenosine triphosphate to adenosine diphosphate and inorganic phosphate during the performance of cellular work.

and regenerated during growth and metabolism. As the name implies, ATP contains three phosphate groups; these are connected to each other linearly as shown in Figure 21. The three black balls represent the phosphate groups, and the polygons represent the remainder of the molecule. Note that the terminal phosphate group is shown to be attached by a coiled spring. The "spring" model is used to suggest how the energy of ATP is delivered. When ATP is utilized, the terminal phosphate group is liberated; that is, the spring "uncoils." This energy release can be thought of as the discharge of an electrical battery: as the energy is used, the battery charge runs down.

Microbes are unable to store appreciable amounts of ATP, and this means that the ATP supply must be constantly regenerated from its component parts, adenosine *di*phosphate and inorganic phosphate. The continual reconstitution of ATP is achieved by means of an "energy dynamo" that operates as shown in Figure 22. Recharging the cellular battery ("coiling the spring") obviously requires energy, and this can be obtained in various ways—for example, from fermentation of sugar or other organic compounds, from aerobic respiration of organic compounds, or from light (in photosynthetic organisms). The unifying principle is that the dynamo operates in the same fashion in all organisms, whatever their idiosyncrasies or however bizarre their ecological niches. The overall efficiency of the

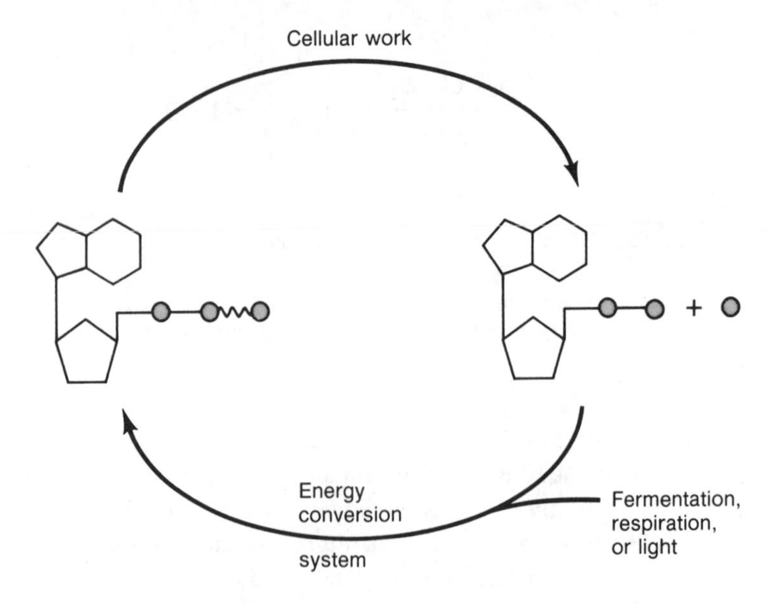

Figure 22 The metabolic energy "dynamo" generates an "ATP current" which is used for performing biosynthetic and other kinds of work.

metabolic energy dynamo varies, however, depending on the kind of recharging process and the ultimate energy source used. Fermentation is the least efficient of the processes noted, but provides a particularly instructive example for analysis.

For each molecule of sugar (glucose) fermented to alcohol and CO_2, a yeast cell gains two ATP molecules in its energy bank. As far as the cell is concerned, this is sufficient for growth, but, even so, yeast is inefficient in its ability to extract the energy potentially available in the glucose molecule. This is true for all anaerobes that are obliged to obtain their energy from fermentation. From the viewpoint of bioenergetics, all (nonphotosynthetic) anaerobes behave in essentially the same way. Because they are inefficient energy converters, they must consume large amounts of sugars (or whatever they ferment) to produce

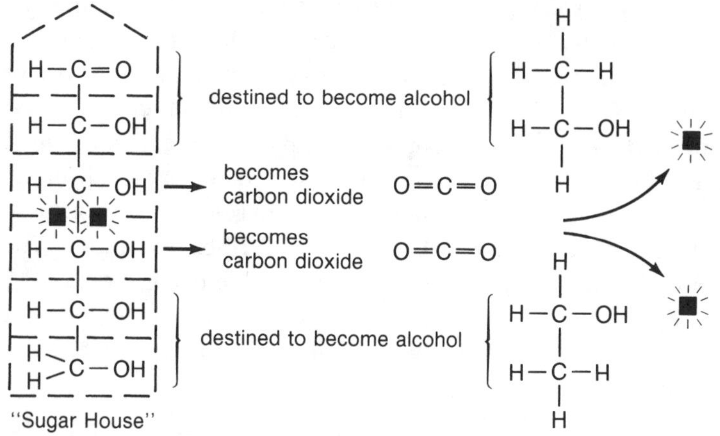

"Sugar House"

destined to become alcohol

becomes carbon dioxide $O=C=O$

becomes carbon dioxide $O=C=O$

destined to become alcohol

Figure 23 Illustration of the fates of the carbon atoms of glucose in alcoholic fermentation. The "sugar house" represents a glucose molecule and the black squares represent energy (ATP molecules) that the yeast cell obtains from the fermentation process.

a relatively meager crop of new cells. Also, they end up swimming in a sea full of energy-rich molecules (such as ethyl alcohol) that they cannot use. In the case of yeast, the energy-rich product of fermentation is alcohol; two molecules of alcohol are formed from each molecule of glucose used.

Glucose, also called dextrose or grape sugar, is the most typical sugar subjected to fermentation. Its six carbon atoms are associated with oxygen and hydrogen atoms ($C_6H_{12}O_6$) linked together in particular ways. In alcoholic fermentation, a series of enzymes acts on the glucose molecule causing progressive alterations, which I will describe in terms of an analogy. Let us suppose that the yeast cell acts as a housewrecker who can get its profit only by obtaining two electrical batteries (representing ATP) concealed in the "sugar house" (representing the glucose molecule) (Figure 23). The black squares in the figure represent the electrical batteries (ATP) which must somehow be extricated.

The yeast cell knows only one way to wreck the house. To get at the batteries, it separates the top three floors from the bottom three. This requires a series of steps; in fact, twelve kinds of machines (corresponding to different enzymes) are needed to pry the "sugar house" apart between floors three and four. The batteries are then used to coil ATP "springs," but the yeast cell is still not finished. It is left with two three-floor apartment units, which must be further dismantled. The final result is summarized on the right side of Figure 23.

From a chemical standpoint, all of the atoms of glucose must be accounted for in the final products. The balance sheet is as follows:

	$C_6H_{12}O_6$ (glucose)	$\rightarrow$	$2C_2H_5OH$ (alcohol)	+	$2CO_2$ (carbon dioxide)
C atoms	6	=	4	+	2
H atoms	12	=	12	+	0
O atoms	6	=	2	+	4

The foregoing discussion was a greatly simplified outline of alcoholic fermentation, but it pinpoints the most essential features. A more conventional, but still fragmentary description of the energy-yielding alcoholic fermentation of sugar would go as follows. The six-carbon sugar molecule is attacked by a series of enzymes that act sequentially. The initial enzyme reactions alter the structure of the sugar molecule, and at a certain stage the modified molecule is cleaved to give two three-carbon molecules. The latter undergo a chemical reaction that yields energy. Rather than being dissipated as heat, however, the energy made available is used for the synthesis of ATP (made from adenosine diphosphate and inorganic phosphate); two ATP molecules are regenerated for each glucose molecule fermented. Finally, the two three-carbon molecules remaining from the energy-yielding reaction noted above are broken down by enzymes, giving rise to two molecules of ethyl alcohol and two of CO_2. This last phase of the fermentation process (formation of alcohol and CO_2) is the final reckoning of the atoms present in the original sugar molecule.

THE ROLE OF SUNLIGHT

The Contribution of Stephen Hales The concept that sunlight is required for the growth of plants was not firmly established until 1845, but it was dimly foreseen by Stephen Hales (1677–1761), the acknowledged founder of plant physiology and one of the most prominent English scientists of the mid-eighteenth century. After study at Cambridge, Hales was ordained and became a minister in the village of Teddington, England. He was indeed a man for all seasons, mixing parish duties with imaginative and bold experiments.

In 1727, Hales published *Vegetable Staticks*, in which he states: "Plants very probably draw through their leaves some part of their nourishment from the air; may not light also be freely entering surfaces of leaves and flowers contribute much to ennobling the principles of vegetables?" It seems that the writer Jonathan Swift was well aware of Hale's ideas, and it is believed that the first part of Swift's description of the Academy of Lagado in *Gulliver's Travels* (1727) was intended to mock Hales.

> This Academy (at Lagado) is not an entire single Building, but a Continuation of several Houses on both Sides of a Street; which growing waste, was purchased and applyed to that Use. I was received very kindly by the Warden, and went for many Days to the Academy. Every Room hath in it one or more Projectors; and I believe I could not be in fewer than five Hundred Rooms.
>
> The first Man I saw was of a meagre Aspect, with sooty Hands and Face, his Hair and Beard long, ragged and singed in several Places. His Clothes, Shirt, and Skin were all of the same Colour. He had been Eight Years upon a Project for extracting Sun-Beams out of Cucumbers, which were to be put into Vials hermetically sealed, and let out to warm the Air in raw inclement

> Summers. He told me, he did not doubt in Eight Years more, that he should be able to supply the Governor's Gardens with Sunshine at a reasonable Rate; but he complained that his stock was low, and intreated me to give him something as an Encouragement to Ingenuity, especially since this had been a very dear Season for Cucumbers. I made him a small Present, for my Lord had furnished me with Money on purpose, because he knew their Practice of begging from all who go to see them.

This fascinating excerpt is an astonishing commentary in several ways. Although Swift obviously considered the project absurd, we find here not only the concept that plants might use light energy, but also the notions that the energy might be stored and even be extracted again. Extraction of the energy, previously obtained from light, is precisely what fermenting yeast cells do. In other words, fermentation of sugar provides yeast cells with energy that was stored in the sugar molecule by the process of photosynthesis. Thus, fermentation is a means of transforming solar energy entrapped by plants into simple energy-rich molecules (alcohol). Large-scale production of fermentation alcohol, which can be used as a fuel for man-made machines, is now being promoted in a number of countries as a means of helping to meet the increasing energy requirements of modern societies.

How Solar Energy is Stored in Sugar The production of the organic matter of plants from CO_2 and water requires a large input of energy, and this is provided by light. Although this is true for all the organic components of plant tissues, biochemists traditionally represent the photosynthetic process in terms of sugar (carbohydrate) formation only. Thus, the familiar equation for green plant photosynthesis is as follows:

$$6CO_2 + 6H_2O \xrightarrow[\text{energy}]{\text{light}} C_6H_{12}O_6 + 6O_2$$

The energy required to make this reaction go is at least 112,000 calories for each 44 grams of CO_2 used. This energy is present in the sugar molecules in the form of chemical bonds between atoms (for example, bonds between carbon and hydrogen atoms, represented as C—H). In other words, the light energy is stored in sugar molecules (which occur in various forms such as cellulose and starch); in the same way, energy is also stored in other organic plant materials such as fats and proteins.

The equation of photosynthesis given above is deceptively simple. The conversion of CO_2 to sugar actually occurs through an intricate mechanism of many steps in which chlorophyll (the light absorber) and numerous enzymes participate. Details of the pathway of carbon transformations in photosynthesis are not particularly relevant here, but one aspect *is* noteworthy: the means by which energy is circulated in the internal workings of photosynthesis. Living cells, whatever their nature, frequently use the same kinds of basic plans for doing things, and as already noted, this applies to biological energetics. Consequently, it is no great surprise to learn that light energy is converted to the chemical energy of ATP, which then becomes the actual driving force for converting CO_2 to sugar. To illustrate the energy dynamics in photosynthesis, we can again use Figure 22. "Cellular work" (at the top of the figure) includes such processes as conversion of CO_2 to sugar.

It is important to note that the mechanisms by which fermentation and light recharge the ATP "battery" are quite different. On the one hand, in fermentation, the energy for ATP resynthesis comes directly from chemical energy already contained within the organic molecule being fermented. On the other hand, in photosynthesis, the energy needed for reforming ATP is delivered by what could be called an electrical system. In effect, light absorbed by membrane-bound chlorophyll creates a current of electrons, which are abstracted from the hydrogen atoms of water molecules; the flow of this "electrical" current through the membrane provides the required energy.

PHOTOSYNTHESIS IN BACTERIA

During the last two decades of the nineteenth century, numerous kinds of bacteria were discovered for the first time, including some pigmented "purple" organisms that seemed to be influenced by light. Eventually it was shown that the purple bacteria could indeed use light as the source of energy for growth. One observation, however, continued to perplex investigators. The purple bacteria, in contrast to green plants, did not appear to produce oxygen gas. An overpowering conviction that the photosynthetic process must be essentially the same wherever it occurs led to repeated attempts, for almost forty years, to demonstrate O_2 formation by the bacteria. Consistently negative results were obtained and finally it was recognized that there must be two major forms of photosynthesis, oxygenic and nonoxygenic. As already noted, the green plant photosynthetic process always results in O_2 formation.

In the *microbial* universe, we encounter both of the two alternative photosynthetic mechanisms. The O_2-producing variety is found in algae and in bacteria called cyanobacteria (sometimes called *blue-greens* because of the color of their pigments) whereas the *non*oxygenic type of photosynthesis is characteristic of bacteria that we will refer to as purple bacteria. All photosynthetic organisms contain a green pigment, one or another type of chlorophyll, which occurs in two forms that absorb light differently (see later). In addition, all photosynthetic organisms contain "accessory pigments," usually of several kinds.* The accessory pigments are usually responsible for determining the apparent color of the organism. Typical cyanobacteria are a bluish-green color, but some species are blackish-green, olive-green, orange-yellow, or reddish-brown. The nonoxygenic *purple bacteria* also show a wide range of colors; purple, purple-

*Accessory pigments of photosynthetic organisms are frequently *carotenoids*. These pigments occur in a number of chemical forms that have yellow to violet colorations. All natural carotenoids are chemically related to a red tomato pigment called lycopene.

red, brown-red, brown-green, yellow-green, etc. Cyanobacteria and purple bacteria show a number of overlapping similarities; thus, they both can use light as the energy source for growth and many species of both can fix molecular nitrogen. But their differences allow us to distinguish two main groups as indicated below.

Cyanobacteria	Purple bacteria
Produce O_2	Do not produce O_2
Grow best with visible light rays	Grow most rapidly with infra-red light (rays not visible to the human eye)
Occur most abundantly in the aerobic biosphere	Prefer appropriate anaerobic ecological niches
Examples: *Anabaena, Nostoc, Oscillatoria, Synechococcus*	Examples: *Chromatium, Heliobacterium, Rhodopseudomonas, Rhodospirillum*

Note that the cyanobacteria, like green plants, grow best with visible light rays. They contain a type of chlorophyll that efficiently absorbs light in the visible portion of the spectrum. The purple bacteria, on the other hand, contain a modified kind of chlorophyll that strongly absorbs infra-red light, rather than the visible rays.

The differences noted obviously have important ecological consequences. The cyanobacteria are commonly found in a large range of habitats exposed to air and sunlight, such as ponds, lakes, and oceans. Although widespread in nature, the cyanobacteria have rather restricted capabilities. Most species grow only as photoautotrophs, that is, with CO_2 as the carbon source and light as the obligatory energy source. Their growth rates are relatively slow, but they are hardy and compete well with other microbes, especially where nutrients are in limited supply.

Purple bacteria, on the other hand, show rapid growth rates, and the metabolic versatility of the known types (more than 60 species) is outstanding. Some have the capacity to obtain growth energy in at least three alternative ways; namely, from light, anaerobic fermentation of sugars (in darkness), or aerobic respiration (in darkness). Virtually all known purple bacteria can use atmospheric N_2 gas as the sole nitrogen source for rapid growth, and their versatility extends to the kinds of carbon sources they can utilize. Thus, in the photosynthetic growth mode, a number of species can use either CO_2 as the carbon source—in which case the metabolic pattern is "photoautotrophic"—or simple organic compounds. Photosynthetic growth on organic compounds is described as "photoheterotrophic," which simply means that light is used only as the source of energy for regenertion of ATP, and organic substances from the growth medium furnish the building blocks for biosynthesis of new cell materials. (In oxygenic photosynthesis, light is the source of energy for ATP synthesis and is also involved in furnishing hydrogen atoms, obtained from water, that are needed for converting CO_2 to organic compounds.)

Given their diverse abilities, it is not surprising that purple bacteria are found in a wide variety of ecological niches. At least ten hitherto unknown species have been discovered since 1980, and there are indications that the N_2-fixing ability of purple bacteria may be of considerable importance in maintaining agricultural productivity in rice paddy fields. Recent reports have described a particularly interesting ecological pattern of purple bacteria populations in a small freshwater lake, Lake Cisó, near Barcelona, Spain. The lake is "fed" by waters that contain large amounts of calcium sulfate (gypsum). The sulfate is converted to hydrogen sulfide (H_2S) by nonphotosynthetic bacteria that grow on organic compounds in the lake sediment (see Chapter 11), and the sulfide diffuses toward the surface of the shallow lake. Lake Cisó is surrounded by a thick barrier of trees and shrubs, and this tends to diminish oxygenation of the top waters by wind action. As a consequence, Lake Cisó is, in essence, an anaerobic lake. Many purple bacteria thrive under these conditions, namely, well-illuminated, nutrient-rich, natural waters that are anaerobic. Thus, the purple bacteria grow very abundantly in the lake, and the surface layers show dramatic color changes. Typically, the water is bright red or brown due to massive accumulations of pigmented photosynthetic bacteria. Occasionally, however, the water becomes clear; in other words, the purple bacteria seem to disappear periodically. This is due, at least in part, to the activities of some newly discovered bacterial species that are parasitic on the purple bacteria! One of these predatory species has been appropriately named *Vampirococcus* (the "coccus" ending simply means the bacteria are spherical in shape).

The similarities and differences between the cyanobacteria and the purple bacteria evoke many questions, particularly about biochemical and cellular evolution. It seems very plausible that these two groups of bacteria are closely related from an evolutionary standpoint. The alternative possibility, that they are not related, would obviously require a remarkable amount of rationalization. Many considerations indicate that the first photosynthetic organisms on Earth were nonoxygenic bacteria that resembled some of the purple bacteria species now extant. Evolutionary changes in some ancient purple bacteria presumably led to the appearance of oxygenic cyanobacteria, which are regarded as the precursors of green plants. Verification of this scenario will require much more research using the diverse approaches of, for example, biochemistry, biogeochemistry, and molecular biology. Without doubt, the isolation and detailed study of hitherto unknown kinds of photosynthetic bacteria will provide the links that are still missing in our comprehension of the evolutionary changes that occurred in early life forms on the Earth.

As far as we know, sunlight is the ultimate source of all biological energy. The term photosynthesis is used to connote the mechanisms used for the conversion of light energy to chemical energy utilizable for growth of cells and organisms. However, many nonphotosynthetic organisms use light in a different way, namely, to obtain information about their environment. The very complex process of vision in vertebrate animals is an important example of the use of light by "photoreceptors" for securing information.

14

The Role of Vitamins

INTRODUCTION

Vitamins deserve special attention for several reasons. They are chemical substances required for normal functions in all cells and organisms, from microbes to humans. If a cell or an organism is unable to manufacture a particular vitamin it needs (for example, vitamin B_{12}) and the vitamin is not available in its diet, the cell or organism will suffer from a deficiency disease which frequently results in death. The remarkable aspect of vitamin function is that only very small quantities are needed for normal growth and maintenance of living cells. Thus, an amount of vitamin B_{12} that is barely visible to the naked eye is sufficient for a large animal for several weeks. How can traces of these chemicals have such huge effects? After biochemists had unraveled the mechanisms that cells use to conduct their chemical processes (metabolism), the explanation of the potency of small doses of vitamins could be easily explained.

VITAMINS AND COENZYMES

The word *vitamin* was coined in 1912 to describe substances that were thought to belong to a category of organic compounds called *amines* that were *vital* for survival of certain microbes and health in humans and various animals. This resulted in the term *vit-amines* or simply "vitamins." It turned out that as more and more vitamins were discovered and characterized, some of them were actually *not* amines, but the name stuck. Vitamins are organic compounds of relatively small size (as compared with macromolecules such as proteins), that are

1. found in foodstuffs in very small quantities;
2. chemically distinct from the main components of foodstuffs; and
3. required in the nutrition of many organisms, including microbes. When absent from the diet of such organisms, a specific deficiency disease or death (or both) results.

As far as is known, all vitamins perform their vital functions in association with particular enzymes that are essential for normal metabolism. Remember that enzymes act catalytically, that is, they accelerate chemical reactions, but remain unchanged themselves. Thus, a small quantity of enzyme can catalyze a relatively large amount of chemical change. Some, however, do not function at all unless combined with a specific "coenzyme," an adjunct of special chemical design. Vitamins form parts of various coenzymes and since these coenzymes are, in turn, parts of (catalytic) enzymes, it follows that vitamins must also act in "catalytic quantities." This explains why vitamins are required in only trace amounts. The principle of vitamin action is illustrated in Figure 24. Once the coenzyme required by a particular enzyme is available, the enzyme can perform its function normally.

The function of vitamins can perhaps be made clearer by a concrete example. The B-vitamin niacin is an excellent case in point. Niacin, also known as nicotinic acid, has the structure:

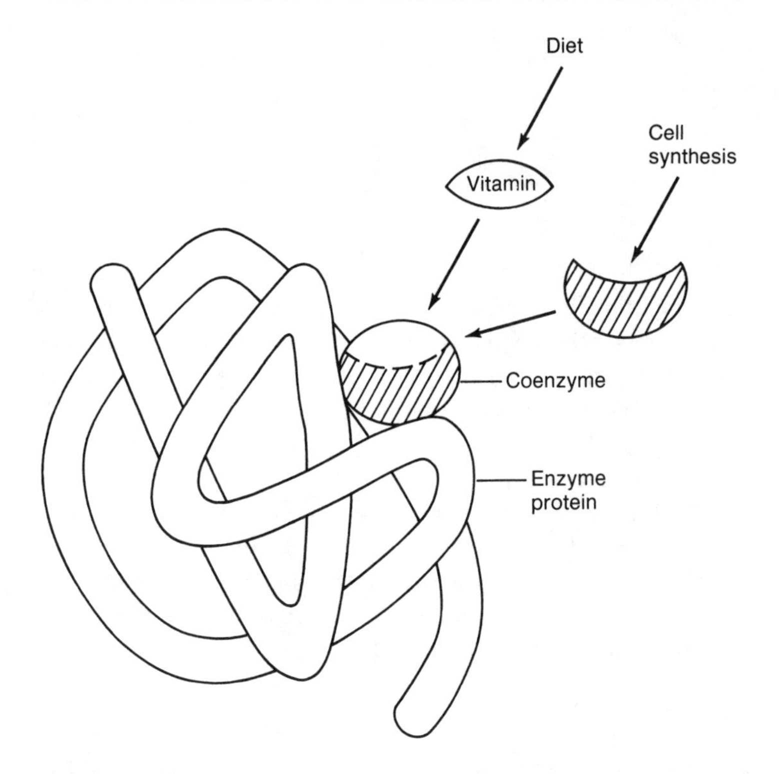

Figure 24 The pathway for combining a vitamin-containing coenzyme with its enzyme protein. The chain of amino acids forming the enzyme protein is represented here as a tubular structure which is folded in a specific way. The relative size of the vitamin-containing coenzyme is deliberately exaggerated; usually a coenzyme is only one-fiftieth to one-hundredth the size of the associated protein.

```
            H
            |
          //C\        //O
    H—C//      C—C
      |        ||     \O—H
    H—C\\     /C—H
          \\N/
```

A simple biochemical reaction converts niacin to niacinamide:

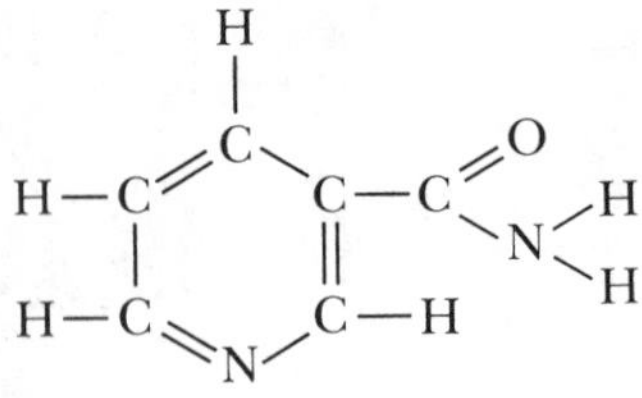

Niacinamide is part of the coenzyme known as nicotinamide adenine dinucleotide, which we designate as NAD (Figure 25). NAD is essential for the activities of a number of enzymes that participate in the energy conversion processes of fermentation and aerobic respiration. A closely related molecule derived directly from NAD functions as a coenzyme in photosynthesis (also an energy conversion process).

The human body is unable to synthesize niacin or niacinamide, and therefore we must obtain the B-vitamin from animal, plant, or microbial foods that we consume. If our diet is deficient in niacin over a long period, the disease condition *pellagra* results (from 1912 to 1916, there were about 10,000

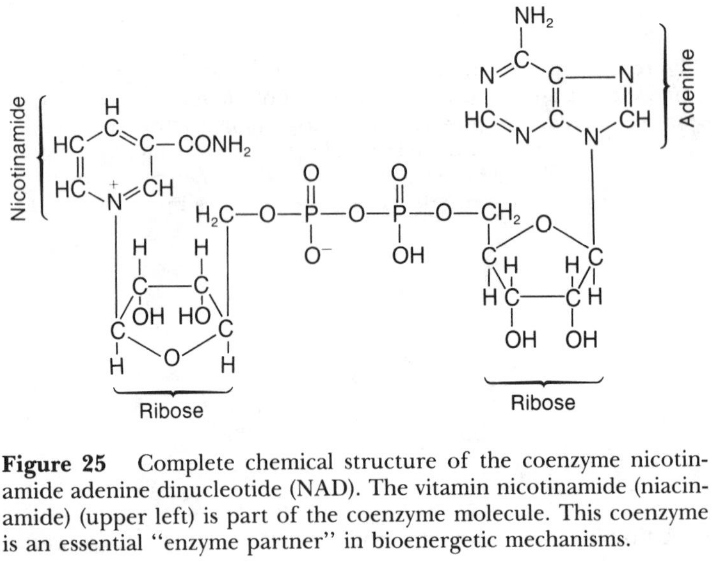

Figure 25 Complete chemical structure of the coenzyme nicotinamide adenine dinucleotide (NAD). The vitamin nicotinamide (niacinamide) (upper left) is part of the coenzyme molecule. This coenzyme is an essential "enzyme partner" in bioenergetic mechanisms.

Table 8 Quantities of the Vitamin Niacin in Some Common Foods

Food	Milligrams of Niacin per 0.25 Pound
Apples	0.1
Cherries	0.3
Peaches	1.0
Carrots	0.6
Corn meal	2.0
Potatoes	1.3
Rice	
Whole	5.2
Polished and cooked	0.4
Bacon	2.0
Lean sirloin	5.7
Pork liver	18.0
Turkey	8.7

deaths each year in the United States due to pellagra; in 1941, the figure was 1900). The recommended daily intake of niacin for the average weight male is 18 milligrams and about 15 milligrams for the average female. Table 8 shows the quantities of niacin present in some typical foods. Long before pellagra was known to be caused by a deficiency of dietary niacin, researchers knew that a "pellagra-preventing factor" was present in different foodstuffs, especially in meats. The "P-P factor," as it was called, was finally isolated and identified as niacin in 1937 by biochemists at the University of Wisconsin.

What about niacin and NAD in microbes? Most microbes can synthesize the niacinamide needed to fabricate NAD from the simple nutrients used to make culture media. But there are some that are unable to do so. If niacin or niacinamide is not furnished to such organisms, they are unable to grow and will promptly die. An example of a bacterium unable to make its own niacin is *Lactobacillus arabinosus*, a bacillus that can be isolated from unpasteurized milk that turns sour. This property of *L. arabinosus* happens to be very useful for determining the amounts of niacin present in different foodstuffs. Thus, the extent of growth of this bacterium in a medium devoid of niacin, but containing a foodstuff extract, gives an accurate measure of the quantity of the vitamin in the food in question. This kind of test is known as the *microbiological assay of vitamins*, and it is a valuable technique for obtaining important nutritional information. Other vitamins can be detected similarly using other species of bacteria.

15

Microbes and Sewage Treatment

It has been claimed that the development of procedures for sewage disposal and water purification has done more to promote public health than all the medical advances made thus far in human history. Although this claim is debatable, the importance of keeping our cities free of disposable waste and providing potable water supplies should not be underestimated. In this chapter we will see that microbes play an essential role in water purification and sewage treatment.

THE SEWAGE PROBLEM

Sewage presents several kinds of community problems. Certain kinds of disease-producing microbes can be transmitted via sewage. The offensive odors of sewage are also an obvious nuisance; the British Parliament was once dismissed for the summer because of the intolerable stench of the Thames river!

Archaeological evidence attests to the concern of many ancient communities with regard to a potable water supply. By

Figure 26 The ancient Romans developed an efficient water supply and disposal system which included the Cloaca Maxima, a huge network of sewers. One of the original outlets can be seen in this print by Giovanni Piranesi, made in 1776.

A.D. 300 the Romans had developed remarkably sophisticated water supply systems that included lengthy aqueducts and the use of lead pipe for distribution of water to public and private facilities (Figure 26). The ancient Romans used about 50 gallons of water per day per capita. The quantity of water delivered to Rome in A.D. 100 was approximately 250,000,000 gallons per day!

After the fall of Rome, the status of water supply systems lapsed, and as populations increased, so did the problems of pollution of water and disposal of wastes. By the mid-nineteenth century, diseases caused by bacteria carried by polluted water (for example, typhoid and cholera) were rampant. Tens of thousands of people died in London during the cholera epidemics in 1847, 1849, and 1852–1854. In 1847, London was the largest city in the world and had an enormous waste disposal problem. The first engineer to make a comprehensive study of metropolitan sewerage needs in an official capacity gave this testimony of the condition of London basements and cellars at that time:*

> There are hundreds, I may say thousands, of houses in this metropolis which have no drainage whatever, and the greater part of them have stinking, overflowing cesspools. And there are also hundreds of streets, courts and alleys that have no sewers; and how the drainage and filth are cleaned away and how the miserable inhabitants live in such places, it is hard to tell.
>
> In pursuance of my duties from time to time, I have visited very many places where filth was lying scattered about the rooms, vaults, cellars, areas, and yards, so thick and so deep that it was hardly possible to move for it. I have also seen in such places human beings living and sleeping in sunk rooms with filth from overflowing cesspools exuding through and running down the walls and over the floors. . . .The effect of the effluvia, stench, and poisonous gases constantly evolving from these foul accumulations were apparent in the haggard, wan, and swarthy countenances and enfeebled limbs of the poor creatures whom I found residing over and amongst these dens of pollution and wretchedness.

Obviously, it was time to revive the water engineering practices of the ancient world and to develop new ways of obtaining the enormous volumes of potable water needed in ruban centers. Public health acts were legislated in England, and by the turn of the century, sewage was collected and large-scale water purification systems had been developed by bacteriologists and engineers. By 1900, most towns in the United States with populations greater than 2000 had adequate water supply systems.

MODERN SEWAGE TREATMENT

The main ultimate purpose of sewage treatment facilities is to treat polluted water so that in the shortest possible time it is

---

*From Metcalf, L., and Eddy, H. P., 1930, *Sewerage and Sewage Disposal*. McGraw-Hill, New York.

converted to "pure" water suitable for human use, or at least pure enough to put into the ocean or a nearby river without polluting it. This is accomplished by intensifying the activities of the assemblage of microbes that normally mineralize organic substances in the natural environment. When a small amount of sewage is dumped into a flowing river, microbes present in the river water and in the sewage grow and multiply. Their biochemical activities eventually result in mineralization of all organic matter, that is, conversion to CO_2, ammonia, nitrate, sulfate, and phosphate. Anaerobes and aerobes are involved at different stages of the process, and the supply of oxygen gas becomes a critical factor. If the amount of oxygen available is too low, foul odors develop due to hydrogen sulfide and noxious organic compounds produced by various microbes under anaerobic conditions. This leads to death of fish and water plants. In contrast, if the amount of sewage added to the river is relatively small and the O_2 supply is sufficient, the water some distance downstream is found to be clear, sparkling, and usable. The self-purification of a river system is due primarily to the metabolic activities of bacteria, and the aim of sewage treatment disposal plants is to accelerate these activities under controlled conditions.*

The magnitude of the water-processing problem is illustrated by an analysis made by John Postgate of the University of Sussex for a typical city in England:†

> In the old West Middlesex area of London the population uses over 50 gallons of water a day per head, all of which washes detritus to the local sewage works. An installation serving one

---

*The catastrophic effect of killing the microbes in a major river with toxic chemicals will no doubt gradually be revealed in France and West Germany. On November 1, 1986, water which had been used to put out a fire in a Swiss chemical plant contaminated the Rhine River. This water contained many tons of poisonous substances which immediately killed more than a half million fish. The death of the microbes that normally purify river waters can be expected to have a number of other undesirable ecological consequences.

†From Postgate, J., 1986, *Microbes and Man*, 2nd ed. Penguin Books, Baltimore.

> and a half million people must handle more than seventy-five million gallons of raw sewage a day, which it collects through a local network of pipes running from drains, sinks, baths, lavatories and industrial effluent conduits (the sewerage system). This sewage represents something like five thousand tons of organic matter; something has to be done with it before it gets into the rivers and seas, or it would cause unimaginable pollution as aquatic microbes recycled its carbon, nitrogen, sulphur, phosphorus and so on. In effect, what a sewage works does is to allow these processes to carry on in controlled conditions, so that the water which carried the sewage is purified and the solid components of sewage are rendered innocuous. This is easily done by modern sewage techniques: the processed solids reach a state in which they can be sold as soil conditioners or fertilizers and the treated water is so pure that, at the Mogden works west of London, for example, the staff will demonstrate the purity of their effluent water by drinking a glass for visitors. (The visitors are unaware that they do something of the sort themselves daily: the water economy of this country is such that quite a lot of purified water finds it way back into the drinking reservoirs. I used to wonder how often an average glass of water had been drunk by someone else before I consumed it. Then I learned from Professor Hutner of New York that London water has passed through an average of seven sets of kidneys when it is drunk. Now I am wondering how the calculation is done.)

Municipal sewage treatment plants appear to be of complicated engineering design, but the principles of the purification process are relatively simple (Figure 27).

The first steps in the purification process consist of screening out large objects and particles and then passing the sewage through sedimentation tanks, to separate it into heavier insolubles and a soluble supernatant fluid.

Anaerobic Decomposition The solids from the sedimentation tanks are passed very slowly through large *anaerobic digestion tanks* for 2 to 4 weeks. Solid matter settles to the bottom of the tanks, and the myriad of anaerobic microbes present grow using organic matter as sources of carbon and energy. This results in breakdown of the larger organic molecules into

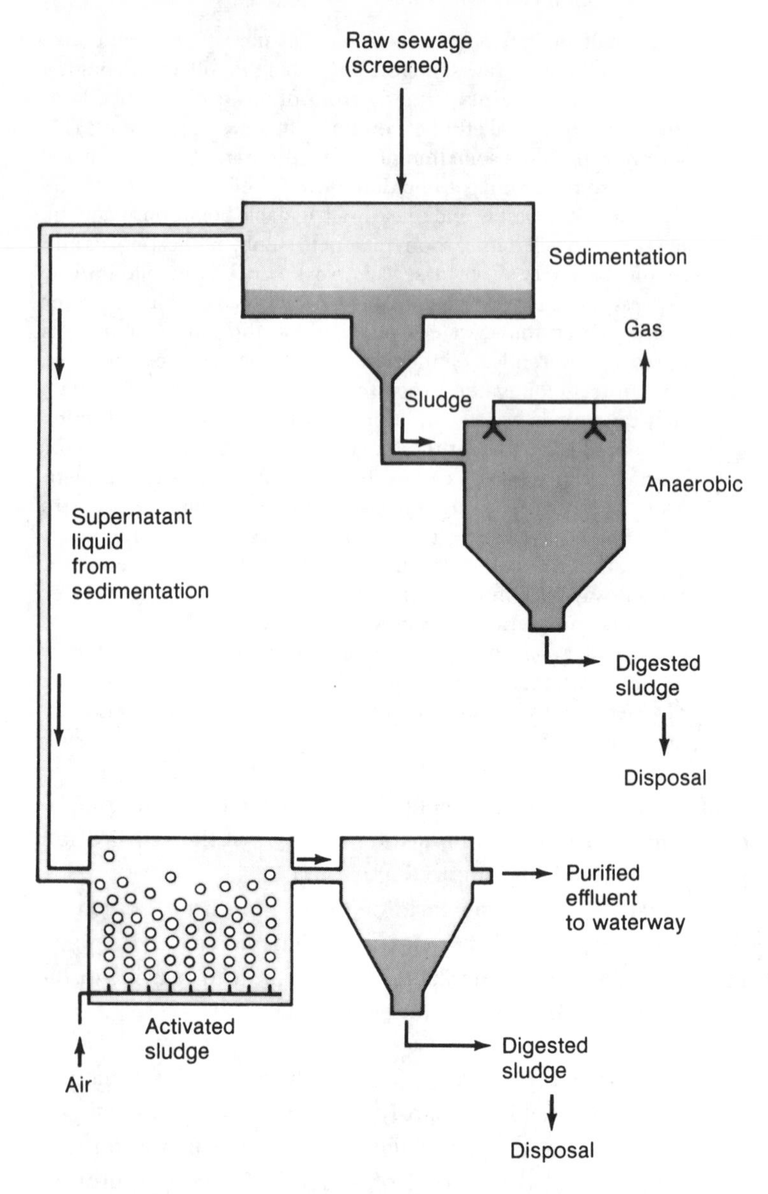

Figure 27 Diagram of a sewage treatment plant.

two components: a comparatively small assortment of small organic molecules, each kind typically containing only two to six carbon atoms, and a variety of gases including carbon dioxide (CO_2) and methane (CH_4), plus traces of hydrogen gas (H_2), ammonia (NH_3), and hydrogen sulfide (H_2S). The gases are bled off and collected. Clearly, we are dealing here with the same circumstances that occur in rumen symbiosis, that is, breakdown of organic materials accompanied by generation of CO_2 and H_2, followed by hydrogenation of CO_2 with H_2, by methanogens, yielding CH_4. Combustion of this bio-gas can be used for heating purposes or running engines and other electrical devices of the sewage plant, as previously mentioned in Chapter 9.

The resulting product, or sludge, consists of indigestible matter and settled bacterial cells and is removed intermittently for disposal (in some instances, it is dried, sterilized, and then used for garden fertilizer).

Aerobic Processes The fluid portion of the sewage emerging from the initial sedimentation process contains large amounts of organic materials, and these are effectively (and rapidly) converted to CO_2 by aerobic bacteria. Several kinds of engineering designs have been employed to accomplish this phase of sewage purification; the *activated sludge process* is now in common use. In this procedure, compressed air or pure oxygen gas is vigorously bubbled through the fluid to hasten bacterial respiration of organic substances.

By the time sewage has been treated in these first two phases, all pathogenic microbes have usually been removed or killed. To be certain that this is indeed the case, modern facilities include a terminal heating step using gigantic pressure cookers. Alternatively, chlorination can be used to ensure destruction of remaining pathogenic organisms. In conventional sewage treatment systems, most of the original nitrogen and phosphorus atoms leave with the final effluent in the forms of nitrate, or ammonia, and phosphate. These inorganic nutrients can cause problems in the waters that receive the sewage plant effluent, for example, by encouraging *eutrophication* (nutrient enrichment of natural waters, frequently leading to excessive growth of algae). There are ways of removing these inorganic nutrients from the final effluent (so-called polishing treatments), but they are not in widespread use as yet.

If There Were No Microbes Chapter 8 began with a brief description of the dismal series of events that would occur if the Earth were to collide with the tail of a comet containing a mysterious gas that could destroy all microbes without doing any damage to plants or animals. The expected consequences of a hypothetical event of this kind were detailed in *Microbes of Merit* by Otto Rahn (see footnote on page 63). The scenario ends with a description of the problems of sewage disposal and water supply in the absence of microorganisms:

> All sewage must ultimately go into rivers, lakes or oceans. As long as we had bacteria, they decomposed the organic matter of sewage slowly, but completely. Some towns dumped all their sewage, after preliminary purification, into a river, and other towns, farther down the river, used that same water for their water supply without fear or loathing because it had become completely purified, thanks to the bacteria. Now that bacteria have completely disappeared from the face of the earth, there is no danger from contagious intestinal diseases, no danger of bad odors from putrefaction, no danger to aquatic life through exhaustion of the oxygen in the water. But all that just indicates that the sewage, after flowing down the river 50 or 100 miles, is still sewage, that means a murky liquid made cloudy by finely suspended particles of human excreta, and containing relatively large amounts of urea. The sewage emptied into the Mississippi at St. Paul arrives weeks later at New Orleans in the same condition, only more concentrated by the sewage from all communities along the river, including Chicago. It is unavoidable that in a short time, all lakes and even the oceans will contain sewage in very noticeable amounts. This sewage does not smell, and people might to some extent overcome the feeling of repulsion for fecal matter, but it does not seem very probable that

swimming in sewage, even in diluted sewage, would be a very popular sport.

As to the drinking water, the solution is simple. Rain water is as pure as ever, and all house owners could construct their roofs so as to catch the largest amount possible, and store it in cisterns. Deep wells may become as precious as oil wells. Our water supply will be smaller, but it will be good.

That is the prospect of life without microbes. It will seem a strange life to us who take the cooperation of microbes for granted, and it will be a hard life, but probably we can make it. Although we will be safe from any contagious diseases for ever, life without microbes may seem hardly worth living to most of us.

Let us hope that we never collide with the tail of such a comet.

16

Plagues and the Origin of the Germ Theory of Disease

Plagues have been the scourge of mankind since time immemorial. The causes of these terrible visitations were unknown for many centuries. In 1835, the first clear-cut evidence was obtained that a microbe was responsible for an infectious disease of an animal. It is remarkable that well before that time, vaccination to prevent infection caused by an unknown entity was being practiced on humans. Obviously, the systematic development of procedures to prevent or ameliorate the effects of pathogenic microbes could not proceed until the basic properties of disease-producing agents were understood. This chapter and the following one deal with the history of communicable disease in the ancient world and with several pioneers of infectious disease research.

DISEASE IN THE ANCIENT WORLD

Throughout the ages, human communities have been subject to the sudden onset of devastating plagues and pestilences. Major episodes are described in considerable detail in numerous records of the past, as far back as 3000 B.C., so it is no surprise that the Bible refers to many different diseases. Leprosy is a particularly good example. The laws for Jewish customs and ceremonies include detailed regulations on how to deal with lepers, as well as their clothes and houses. From Leviticus:

> When leprosy is probable the priest shall look on him and pronounce him unclean, and he shall be shut up and inspected at the end of seven days. If the disease has not progressed the patient shall be shut up for another seven days and examined again, and if the plague be dim and the plague be not spread in the skin the priest shall pronounce him clean, and he shall wash his clothes and be clean. . .if the disease has spread the patient is pronounced to be leprous, and the leper in whom the plague is, his clothes shall be rent and the hair of his head shall go loose, and he shall cover his upper lip and shall cry, Unclean, unclean. All the days wherein the plague is in him he shall be unclean; he is unclean: he shall dwell alone: without the camp shall his dwelling be.

Imagine now that you are a citizen of the democracy of Athens in 430 B.C. It is the "golden age" of Pericles, and Athenians have a high standard of living. The Great Peloponnesian War is underway, and a fleet of 100 Athenian ships has sailed along the southern peninsula of Greece raiding the inhabitants. One day in overcrowded Athens, a plague begins and spreads rapidly. Victims develop a raging fever, an extreme thirst, and a bloody tongue and throat. Pustules and ulcers break out on the skin and, finally, the victims die. The plague also strikes aboard ships of the fleet, forcing their return to Athens. Before the plague subsides, at least one-fourth of the fighting men and one-third to two-thirds of the total population are dead. The morale of those still living is shattered. Thucydides, a Greek historian of the time, described the demoralization as follows:*". . .fear of gods or law of men there was none to restrain

---

*From Cartwright, F. F., 1972, *Disease and History*. T. Y. Crowell, New York.

them. As for the first, they judged it to be just the same whether they worshipped them or not, as they saw all alike perishing; and as for the latter, no one expected to live to be brought to trial for his offences."

Reliable historical accounts detail the ravages of pestilences and plagues over the past two thousand years. In A.D. 166 Roman legions returning from Syria brought back a pestilence that spread throughout the countryside, eventually reaching Rome; corpses were carried away from the city by the wagon-load. The course of human history was significantly affected by such episodes, including three major pandemics (related epidemics occurring over a large area): one in A.D. 540 to 590; the Black Death of 1346 to 1361, which wiped out about one-fourth of Europe's population; and the Great Plagues of 1665 to 1666.*

It was not known what was causing these terrible devastations. Some thought they were divine judgments to punish the wickedness of mankind and others thought that some sacrifices would help to appease the anger of the gods. Some people died, and others did not, rich and poor and good and bad; it seemed to make no sense. Nevertheless, it gradually became clear that in an epidemic, your chances might be better if you avoided victims and if the dead were disposed of quickly (preferably by burning) with a minimum of contact.

Since the nature of these disasters was not understood, all kinds of fallacious ideas and practices were prevalent. One example is of interest in this connection. Early in the seventeenth century a special costume for doctors became popular in Italy and France, a robe made of fine linen cloth coated with a paste of wax in which aromatic substances were incorporated. The wax surface was supposed to prevent "plague miasmas" from sticking to the smooth slippery surface. In Italy, the robe was

---

*For detailed accounts of how epidemics have affected world history (including how typhus helped to "defeat" Napoleon), see the reference on page 145 and Zinsser, H., 1935, *Rats, Lice and History*. Little, Brown, Boston.

frequently topped with a hood, fitted with a large and sinister beak-like nose that was filled with materials saturated with perfumes and alleged disinfectants. It was a popular belief that strong-smelling substances could prevent disease—so people burned tar, old shoes, and the like in their homes.

THE BEGINNING OF AN ANSWER

William Boghurst, a pharmacist, was one of the heroes of the Great Plague of London (1665). He remained in the city to help the miserable victims (during September 1665, more than 7000 deaths occurred each week). The following excerpt describes Boghurst's activities:*

> [He] commonly dressed forty Plague sores a day, and in diagnosis would test the pulse of a patient, sweating in bed, for five and six minutes. He upheld in their beds those threatened by strangling and choking, often for half an hour together, the breath frequently falling on his face. In the haste of a busy day he would eat and drink with the Plague-stricken, sitting on the edge of the bed and talking with them, often watching the death and closing the mouth and eyes—for in death commonly the mouth was wide open and eyes staring. Help being scarce in the infected houses, he at times assisted to lay out the corpse and afterwards place it in the coffin, and as a last act of charity he might accompany it to the grave.

The best explanation Boghurst could offer for the cause of the disease was that†

> Plague or pestilence is a most subtle, peculiar, insinuating, venemous, deleterious exhalation arising from the maturation of the faeces of the earth extracted in the aire by the heat of the

---

*From Bell, W. G., 1951, *The Great Plague in London in 1665*, revised ed. The Bodley Head, London, England.

†From Bulloch, W., 1938, *The History of Bacteriology*. Oxford University Press, Oxford, England.

> sun and difflated from place to place by the winds and most tymes gradually but sometymes immediately agressing apt bodyes.

The plagues and pestilences were, of course, caused by microscopic parasites such as bacteria, fungi, and other "invisible" biological agents. Before 1674 the existence of invisible microbes was undreamt of, and the idea would have been considered a fantastic notion to most people. How then did the germ theory of infectious disease originate and develop? The usual, and erroneous, answer is that the theory was the creation of Louis Pasteur. In fact, he became interested in the problem long after the concept was first proposed and evidence supporting it obtained.

Students of the history of biological science are taught that the first really perceptive insights into the nature of infectious disease were advanced by the Italian Girolamo Fracastoro (ca. 1478–1553). Fracastoro was a physician, astronomer, geographer, poet, and humanist. He lived for some time in Verona, Italy (witnessing a great epidemic of plague there), and later settled in a villa on the shore of Lake Garda for a life of contemplation and study. Fracastoro was an acute observer and published an important early work on syphilis. Other books (published in Latin) dealt with the essence of contagion. He spoke of the "seminaria" of disease; the word is translated as "seeds" or "germs," and some scholars believe that he considered them to be living entities. It was clear to Fracastoro that there were several kinds of contagions. For example, in one category, the causative agent is transferred by touching contaminated clothing, wooden objects, and the like. In another category, the contagion appears to be transferrable only by direct contact between individuals. He also realized that different diseases showed different patterns; for example, some preferentially attacked children and others particularly affected certain organs of the body. Remarkably, his analyses of epidemics of plague, typhus, syphilis, and other infectious diseases came close to explaining their true nature. By the end of the sixteenth century, however, his work had been forgotten, mainly because of the lack of scientific communication during the seventeenth century.

In 1835, Agostino Bassi (1773–1856) published the first definitive evidence for microbial causation of an infectious disease in animals, in the form of a monograph. Bassi was an Italian lawyer and naturalist who abandoned public posts in 1816 to devote full time to agriculture. At the time, *muscardine*, a disease of silkworms, was ravaging the silkworm industries in Italy and France. The prevailing notion was that death of the worms was due to some vague environmental cause (state of the atmosphere?). Bassi had the idea that the disease was caused by an "extraneous germ," and he soon discovered that a white material which always developed on dead worms was the infectious matter. He concluded that every outbreak of the disease could be traced to infected silkworms or use of contaminated cages or utensils. Moreover, he demonstrated that suitable precautions could prevent outbreak of the disease, for example, disinfection of silkworm eggs with alcohol, and disinfection of all instruments and implements used in the nursery. In 1834, after many years of work on the subject, Bassi performed a series of experiments before a commission from the Faculties of Medicine and Philosophy, University of Pavia, Italy, with the purpose of communicating his findings and saving the silk industry. The commission members issued a signed certification which was reprinted in the preface of Bassi's 1835 monograph *Del mal del segno* (Figure 28):*

> Signor Doctor Bassi of Lodi in 1833 applied to the Imperial Royal University of Pavia for permission to communicate some of his experiments and findings on the disease of the silkworm called *il segno*. But because during that year the appropriate

*English translation by Yarrow, P. J., 1958, *Phytopathological Classics*, vol. 10. American Phytopathological Society, Baltimore.

(a)

Figure 28 (a) Founder of the germ theory of infectious disease, Agostino Bassi. (b, *next page*) The title page of Bassi's 1835 publication on the muscardine disease of silkworms, in which he identified a fungus as the causative infectious agent. He suggested that the disease could be eliminated by preventing microbial contamination of worms, disinfection of worm-breeding rooms, and sterilization of implements and equipment. Disinfectants suggested by Bassi included: caustic potash lye, nitric acid, and spirits of wine or crude brandy.

experiments could not take place, he renewed his application during the current year, 1834. He conducted the experiments in the presence of a Commission composed of members of the faculties of Medicine and Philosophy, which reached the following conclusions:

1. The white substance, crust, or efflorescence on the silkworm is indeed infectious, and hence placed in contact with a healthy insect will transmit and propagate the disease.
2. The efficacy of this substance can be destroyed by various chemical agents which do not damage the insect. This can

DEL MAL DEL SEGNO
CALCINACCIO o MOSCARDINO
Malattia che affligge
I BACHI DA SETA
E SUL MODO
DI LIBERARNE LE BIGATTAJE
ANCHE LE PIÙ INFESTATE
Opera
DEL DOTTORE AGOSTINO BASSI
DI LODI
la quale oltre a contenere molti utili precetti intorno al miglior governo dei Filugelli, tratta altresì delle Malattie
DEL NEGRONE E DEL GIALLUME

LODI
DALLA TIPOGRAFIA ORCESI
1835

(b)

be done before the said substance is brought into contact with the insect or after, provided the remedy is applied soon after contamination.

3. In view of the extreme ease with which this infectious substance spreads, and adheres to everything firmly; and considering the minute size of its particles in consequence of which a single dead worm when reduced to the state of efflorescence can infect a whole silkworm nursery, it cannot be doubted that the said substance is the usual cause of the mentioned disease.

4. Seeing that there are chemical agents that can decompose and destroy the infectious substance, the Commission declares its conviction that by the proper use of these agents the all too easy transmission of the disease can be stopped and the disease cured and prevented.

Despite his failing eyesight, Bassi identified the culprit of muscardine as a microscopic fungus. It seemed to be an organism known as *Botrytis paradoxa*. In 1835, an Italian botanist confirmed the identification and renamed the fungus *Botrytis bassiana*. Because of the onset of blindness, Bassi could no longer continue microscopic work, but pursued development of his "parasite theory of disease" in connection with plague, syphilis, cholera, and other infectious processes. This pioneer received a number of awards from both Italian and foreign academies, but his momentous research was not properly appreciated by a number of subsequent investigators, including those of the "Pasteur school."

Earlier in this chapter the term plague was used in the sense of "a great calamity." *The* plague is an infectious disease that can take several forms, depending on the properties of the particular strain of the causative microbe, now known as *Yersinia pestis*. If the bacteria localize in the lung, the disease is called pneumonic plague. The most common form of the disease, bubonic plague, takes its name from the term *bubo*, a hard and painful swelling of a lymphatic gland (which are located in the armpits, neck, and groin). In 1894, Alexandre Yersin (1863–1943), a medical officer in the French colonial service, investigated a bubonic plague epidemic rampant in China. In a small laboratory in Hong Kong, he was able to isolate the plague microbe. The bacterium is harbored by infected rats and is usually transmitted from rat-to-rat and rat-to-man by fleas. The pneumonic form, however, can be transmitted from person to person by contaminated droplets exhaled in the breath.

17

Three Giants of Infectious Disease Research: Pasteur, Koch, and Jenner

Animals that were not seen to be directly born by natural reproduction were considered by the ancients to come into the world through *spontaneous generation* as the result of the combined action of heat, water, air, and putrefaction. J. B. van Helmont (1577–1644), a Belgian alchemist, physician, and philosopher is often quoted as follows:*

> If a foul shirt be pressed together within the mouth of a Vessel, wherein Wheat is, within a few dayes (to wit, 21) a *ferment* being drawn from the shirt, and changed by the odour of the grain, the Wheat it self being incrusted in its own skin, transchangeth into Mice. . . .And which is more wonderfull, out of the Bread-corn, and the shirt, do leap forth, not indeed little, or sucking,

---

*van Helmont, J. B., 1662, *Oriatrike or, Physick Refined.* Printed for Lodowick Loyd, London.

or very small, or abortive Mice: but those that are wholly or fully formed.

CONTRIBUTIONS FROM PASTEUR AND KOCH

Numerous natural philosophers and scientists were occupied with the question of spontaneous generation for centuries, and it was still a hot topic when Louis Pasteur came onto the scene in the 1870s. By that time, it was clear that spontaneous generation of mice, maggots, etc., was unprovable and extremely dubious, and the argument then shifted to microbes. Pasteur took up the challenge and through careful and cleverly designed experiments, demolished virtually all claims made by others for demonstration of spontaneous generation of microbes. About van Helmont's recipe, Pasteur stated, "What this proves is that to do experiments is easy; but to do them well is not easy." The final blows that ended discussion of spontaneous generation were delivered by John Tyndall (1820–1893), an English physicist who became a professor at the Royal Institution in London. According to Bulloch:*

> Whereas the personality of Pasteur inspired something of the nature of opposition, Tyndall's magnetic personality, his exact technical methods, the logic of his interpretations, and the clarity of his literary compositions were acceptable to a large number of intelligent people. The doctrine of the germ theory of disease was then securing a foothold, and Tyndall was one of its early and staunchest upholders. The medical profession owes a debt to Tyndall, and this was partly acknowledged when he was made honorary Doctor of Medicine by the University of Tubingen.

Pasteur's research on the spontaneous generation question was closely related to his work on fermentation, and he later

---

*From Bulloch, W., 1938, *The History of Bacteriology*. Oxford University Press, Oxford.

undertook the study of diseases of silkworms. Emile Duclaux described how this came about. Duclaux (1840–1904) was a French chemist and bacteriologist who served as a professor at several French universities and succeeded Pasteur as Director of the Pasteur Institute in Paris. Duclaux's biography of Pasteur *Pasteur—History of a Mind* was published in 1896, one year after Pasteur's death (English translation by E. F. Smith and F. Hedges, published in 1920 by W. B. Saunders, Philadelphia). According to Duclaux, during a protracted epidemic that was affecting silkworms and ruining the French silk industry, Senator J. B. Dumas convinced Pasteur to work on the problem. Curiously, Duclaux discusses the background of information that was available on silkworm diseases, without mentioning Bassi. It is ironic that Pasteur's misinterpretation of certain observations led him temporarily astray as to whether or not one of the apparently infectious diseases of silkworms was in fact caused by a microbe. In any event, this research was Pasteur's introduction to later study of infectious disease in more highly evolved domestic animals and humans. From 1875 to 1890, Pasteur (a French chemist) (Figure 29) and Robert Koch (a German physician) were in the limelight of research on infectious diseases, and they became fierce competitors in seeking recognition for their discoveries. Unfortunately, their interactions became acrimonious and had nationalistic overtones. Pasteur had remarkable insights and imagination, excellent technical skills, good organizational ability, and political acumen, and in addition, was a formidable warrior. A biography* published three years after his death gives the essence of his disposition:

> It is in this connection that we realize that Pasteur was not only a *savant* content to seek the truth and find it, but that when he had in any matter succeeded in the difficult task of convincing himself, he was impelled with almost a fanatic's zeal to force his conviction on the world, nor did he put up his sword until every

---

*Frankland, P., and Frankland, Mrs. P., 1898, *Pasteur*. Cassell, London.

Figure 29 Commemoration of Louis Pasteur on the French five franc note. *Top*: Pasteur and the Pasteur Institute, Paris. *Bottom*: reverse of the note showing Pasteur and some of the equipment used in his classic research. Also, at left is a depiction of Joseph Meister being bitten by a rabid dog. At the age of nine, Meister was brought to Pasteur in 1885 for treatment of the dog bites. He was the first person to be treated by the new method Pasteur had developed for immunization against rabies. After a series of 13 inoculations, Meister recovered and remained immune.

redoubt of unbelief had been taken, every opponent converted or slain.

Koch, 21 years younger than Pasteur, was also combative, and in some ways excelled Pasteur as an experimenter, at least in bacteriology. His development of novel and important experimental techniques has already been noted in Chapter 5, and these were landmark exploits. Koch's initial research (while practicing as a country doctor) was concerned with anthrax, primarily a disease of cattle, sheep, and horses, but which can also affect other domestic animals and humans. At that time, anthrax epidemics were commonplace in Europe and had ruinous effects on small farms. Koch isolated the bacterium *Bacillus anthracis* from diseased animals in pure culture and showed by the most rigorous criteria that this organism is the causative agent of anthrax. This was the first instance in which a specific microbe was demonstrated to be the cause of an infectious disease in a higher animal. Koch later isolated the bacteria that cause tuberculosis (1882) and cholera (1883). In his research, he refined the strategy required to unambiguously identify the cause of microbial disease, the so-called Koch's Postulates. One version of these is as follows:

1. The microbe must be present in every case of the disease.
2. It must be isolated from the diseased host and grown in pure culture.
3. The same specific disease must result when a pure culture of the microbe is inoculated into a healthy susceptible host.
4. The microbe must be recoverable once again from the experimentally infected host.

These criteria proved to be important in much later research, but in some instances they could not all be met easily. An outstanding example in this connection is leprosy. The bacterium responsible, *Mycobacterium leprae*, was isolated in 1872, but a susceptible laboratory animal was not discovered until 1971 (surprisingly, the armadillo).

VACCINATION AND IMMUNITY

The last phase of Pasteur's meteoric career was concerned primarily with prophylaxis against infectious disease, in particular by *vaccination* procedures. This was not a new concept; inoculation to procure immunity to smallpox had been practiced for centuries. According to F. F. Cartwright,* physicians in ancient China "removed scales from the drying pustules of a person suffering from mild smallpox, ground the scales to a fine powder, and blew a few grains of this into the nostrils of the person to be protected." Another procedure was publicized in 1717 by a remarkable 29 year-old woman, Lady Mary Montagu, wife of the British ambassador to Turkey. She observed that every September a group of old women made rounds of houses in Constantinople, where families would gather for "ingrafting" ("inoculation parties"). Each practitioner carried, in a nutshell, a small sample of pus collected from a victim of a mild attack of smallpox. She would quickly scratch open a vein on a limb of the "customer" with a needle, dip the needle into the pus, smear it on the open vein, and then bind the wound. Lady Mary wrote to a friend about the response of children treated in this way:

> . . .they play together all the rest of the day, and are in perfect health to the eighth. Then the fever begins to seize them, and they keep their beds two days, very seldom three. . .and in eight days' time they are as well as before their illness. . . .Every year thousands undergo this operation; and the French embassador says pleasantly, that they take the smallpox here by way of diversion, as they take the waters in other countries. There is no example of anyone that has died in it; and you may believe I am

---

*From Cartwright, F. F., 1972, *Disease and History*. T. Y. Crowell, New York.

> very well satisfied of the safety of this experiment, since I intend to try it on my dear little son.

Which she did. Her son became the first known Englishman to be vaccinated against smallpox. By 1722, King George I was persuaded to have two of his grandchildren similarly inoculated (beforehand, six prisoners under sentence of death volunteered to be guinea pigs on promise of reprieve). Lady Montagu became a celebrity.

The inoculation procedure worked well most of the time, but there were occasional failures. The method was totally empirical, and sometimes the child would actually become ill with smallpox. This happened to the young Edward Jenner (1749–1823) during a severe epidemic in England. He recovered and was thereafter immune to the disease, which became a definite advantage in his later work. Jenner became a country doctor and used the procedure himself on children of his patients. Jenner was aware of the old wives' tales that people who suffered from the mild disease "cowpox" became resistant to smallpox. Cowpox would first appear on the teats of infected cows as inflamed pustules and would quickly spread throughout the herd. Dairymaids and milkmen would then develop sores on the ends of their fingers and at the finger joints. The sores would spread to other parts of the body and a fever would set in, which usually subsided after a few days.

Jenner hypothesized that cowpox was a form of smallpox, and he closely observed numerous cases. In May 1796, he performed one of the classic experiments in the history of medicine. In his words: "I selected a healthy boy, about eight years old, for the purpose of inoculation for the Cow Pox. The matter was taken from a sore on the hand of a dairymaid who was infected by her master's cows." Jenner smeared pus into several deep scratches on the arm of James Phipps. Seven days later the boy had an eruption on his arm at the site of the scratches and discomfort in his armpits, but he recovered within a few days. On July 1, Jenner inoculated James with "matter" from the pustules of a person ill with smallpox. The smallpox matter had no effect, and Phipps was subsequently inoculated many times in the same fashion with no ill effect. Jenner had reinvented vaccination *as a scientific procedure*.

The causative agents of cowpox and smallpox are *viruses*, not true microbes (this distinction is explained in Chapter 19). The two viruses are closely related, and development of immunity to cowpox also confers immunity to the smallpox virus. Whether the pathogenic agent is a virus or a microbe, the basic mechanisms by which vaccination with some form of the agent gives rise to immunity are the same.

18

Mechanisms of Immunity

INTRODUCTION

It was recognized long ago that in devastating epidemics of infectious disease, some individuals become ill and recover whereas many of the afflicted die. This suggests that in some individuals, there must be natural defense mechanisms that can overcome a microbial (or viral) invader. Indeed, if there were not and if we had no other means of "antimicrobial warfare," the lives of most humans would be a succession of microbial diseases from cradle to grave. We are constantly exposed to microbes in our environment, and almost every accessible surface of the body harbors a large population of diverse bacteria and other microbes. Most are harmless, but some have the potential of causing disease processes. Whether or not disease develops depends on the balance of a number of factors.

The nature of the natural defenses of the body have been under study for over a century, and we now have a good, but still incomplete knowledge of how they work. Prominent among the defenses are special kinds of white blood cells that can engulf and destroy microbes, and the molecular immune system consisting of special molecules called *antibodies* that can combine with and inactivate the pathogenic invader.

But what can be done for individuals whose natural defenses can be breached, and possibly overwhelmed, by a pathogen? One of the truly great contributions of microbiologists to medical practice and public health was the discovery that some microbes produce *antibiotics,* chemicals that can kill other microbes but which do not affect human cells or tissues. Penicillin, produced by a mold, was the first to be discovered, and we now have a fairly large (and growing) collection of such agents. Regrettably, we still do not have effective antibiotics for viral invaders, but we can expect that ongoing research will eventually make them available.

DEFENSE AGAINST MICROBES

Pathogenic microbes gain entrance to the body in characteristic ways, depending on the microbe. The portal of entry for the bacteria that cause typhoid and paratyphoid fevers, dysentery, and cholera is the digestive tract. These pathogens can withstand the enzymes in saliva and other digestive juices, as well as the acidity of the stomach. Certain other microbes, in contrast, enter by way of the respiratory passages, the urogenital tract, or breaks in the skin. As invasive microbes grow in the body, they destroy host cells and tissues by producing toxic substances (*toxins*) and/or special enzymes that attack major components of cell structures. Animals generally possess a series of defenses against microbial invaders.

Primary Defense Aside from the mechanical barrier of intact skin, secretions of the skin contain chemicals that inhibit or kill bacteria. Hairs in the nasal air passages also represent a mechanical barrier; they filter out particles with attached bacteria.

Microbes that enter the eye are subject to being flushed out by eye secretions which contain lysozyme, an enzyme that attacks the cell walls of many bacteria, causing them to lyse (fall apart). Saliva also contains lysozyme. Stomach contents have a low pH (strong acidity), and this quickly kills many, but not all, microbes.

Secondary Cellular Defense An important aspect of defense against microbial infection is the activity of special *phagocytic* cells that are very mobile and can engulf and destroy microbes. There are several kinds of phagocytes widely dispersed in the body, for example, in blood, spleen, liver, and bone marrow. They are varieties of white blood cells that can perform a "seek and destroy" function called *phagocytosis*. Phagocytes were discovered by Élie Metchnikoff (1845–1916), who visualized the animal body as a battlefield fought over by warring microbes and protective phagocytes.

In 1882, Metchnikoff was forced to resign from the faculty of the University of Odessa because of his radical political views. His wife had an inheritance from her parents, so they decided to move to a town on the shore of the Mediterranean where he could pursue his research on zoological marine specimens. It is rare for scientists to remember the exact moment of dramatically new insights, and Metchnikoff's recollections are of interest in this connection:

> I was resting from the shock of the events which provoked my resignation from the University and indulging enthusiastically in researches in the splendid setting of the Straits of Messina.
>
> One day when the whole family had gone to the circus to see some extraordinary performing apes, I remained alone with my microscope, observing the life in the mobile cells of a transparent star-fish larva, when a new thought suddenly flashed across my brain. It struck me that similar cells might serve in the defense of the organism against intruders. Feeling that there was in this something of surpassing interest, I felt so excited that I began striding up and down the room and even went to the seashore to collect my thoughts.

> I said to myself that, if my supposition was true, a splinter introduced into the body of a star-fish larva, devoid of blood vessels or of a nervous system, should soon be surrounded by mobile cells as is to be observed in a man who runs a splinter into his finger. This was no sooner said than done.
>
> There was a small garden to our dwelling, in which we had a few days previously organised a 'Christmas tree' for the children on a little tangerine tree; I fetched from it a few rose thorns and introduced them at once under the skin of some beautiful star-fish larvae as transparent as water.
>
> I was too excited to sleep that night in the expectation of the result of my experiment, and very early the next morning I ascertained that it had fully succeeded.
>
> That experiment formed the basis of the phagocyte theory, to the development of which I devoted the next twenty-five years of my life.

Metchnikoff's assessment of the great importance of phagocytes in disposing of microbes that invade the body has been supported by intensive research over the past few decades. The phenomenon of engulfment and subsequent destruction of microbes by phagocytes is now known to be a complicated multistep process. One indication of this complexity is that if microbes are coated with antibody molecules (see below), the phagocytosis defense is effectively triggered. Individuals with phagocytes that do not function properly may have recurrent infections caused by microbes that are normally not pathogenic, and some forms of "phagocyte cell disease" are often fatal in childhood.

Third Line of Defense: The Immune System Under natural conditions, individuals who recover from illness due to an invasive microbe or virus have special protein molecules in their blood serum that are able to bind tightly to the pathogenic agent. These protein molecules are called *antibodies*, and if their concentration in blood serum remains sufficiently high, the individual will usually be protected against subsequent exposures to the same pathogen. In this phenomenon, the invasive microbe or virus is referred to as the *antigen* (entity that stimulates generation of an antibody). Neutralization of the antigen by the antibody is an important aspect of the immune response. Without question, the exact mechanism by which a particular antigen evokes the formation of antibody molecules that are designed to precisely "fit" with the antigen and thus neutralize it is of extraordinary complexity. Cracking this scientific puzzle has kept large numbers of scientists busy for half a century, and although much progress has been made, many aspects are still not completely understood.

Pasteur's research on vaccination (in connection with anthrax, chicken cholera, and rabies) was concerned with devising procedures for attenuating virulent microbes or viruses so that they would lose their toxic and disease-producing properties, but retain their ability to stimulate the animal body to become immune—in other words, to retain their antigenic capacities. Pasteur was attempting (sometimes successfully) to "clean up" the antigen (the microbe or virus) so that it *selectively* lost the capacity to induce disease processes. It was an empirical hit-or-miss proposition and usually involved simple procedures (for example, mild heating). Simple procedures are still often used in the twentieth century: for example, the Salk polio vaccine was nothing more than poliomyelitis virus treated with dilute formaldehyde for a certain length of time under well-defined conditions. The virus loses its pathogenicity, but still functions as an antigen.

The antigenic part of a microbe or virus is ordinarily (but not necessarily) one or more proteins that make up an integral surface portion of the pathogen. If biochemists can isolate antigenic proteins in pure form, they can be used for eliciting production of the corresponding antibodies by appropriate animal cell systems grown in flasks in the laboratory. Much current effort in biotechnology is being devoted to development of such procedures. In the living animal, antibody formation involves several classes of special white blood cells that originate in bone marrow and in the thymus gland (B-lymphocytes and T-lymphocytes). These cells associate with the antigen in a complicated scenario that results in extensive proliferation of a particular line (clone) of B-cells committed to production of a specific kind of antibody. It is believed that a normal animal can respond to several million different antigens in this way! This astonishing fact means that if any one antigen is represented as a key, the body is capable of producing millions of kinds of locks, each kind being exactly complementary to a particular key. The key–lock analogy has been used for over 70 years to explain the specificity of antigen–antibody combinations (Figure 30). It is noteworthy that AIDS (acquired immune deficiency syndrome) is caused by a virus that has the unique property of destroying the ability of antibody-producing cells to function normally.

A final note on infectious disease: whether or not a human or animal succumbs to infection depends on the net balance of numerous factors, including how well (and how fast) the immune system operates. If the parasite is particularly virulent and the number of invaders gaining entry is large, this can tip the balance of resistance versus infection toward disease. The factors involved in the delicate balance are shown in Figure 31.

MICROBIAL ECOLOGY OF ANIMALS

Thus far we have considered the mouth and intestinal tract of higher animals as locales in which microbes normally occur in large numbers. Theodor Rosebury made noteworthy studies on this aspect of microbiology* and summarizes the overall microbial ecology of humans as follows:

> The life on man consists of microbes, microbes in extraordinary variety and in large numbers. I once counted some 80 distin-

---

*Rosebury, T., 1969, *Life on Man*. Viking Press, New York.

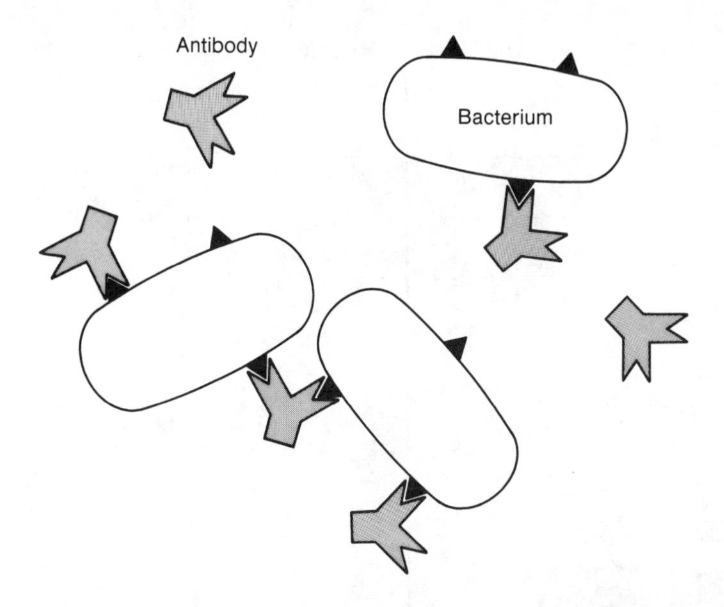

Figure 30 A schematic representation of one kind of antigen-antibody combination. The large rod-shaped bodies represent pathogenic bacteria; the dark protrusions (triangles) on the bacterial surfaces are antigens. The Y-shaped objects are antibody proteins that were produced by the animal after an exposure to the bacterial antigens. Subsequently, antibodies in the blood combine with the bacteria in a very specific fashion, which is analogous to the way a key fits into its complementary lock. As a result of this combination, the bacteria are inactivated. This diagram does not accurately depict the relative sizes of bacterial cells and antibody proteins; in actuality, bacteria are much larger than antibodies (see Figure 33 for actual sizes).

guishable kinds in the mouth, and the total number of bacteria excreted in feces by an adult each day ranges under normal conditions from 10^{11} to 10^{14}—from 100 billion to 100 trillion. The microbes of our normal population inhabit nearly every surface that is freely exposed, such as the skin, or accessible from the outside, such as the lining of the intestinal tract. Along the length of the alimentary canal, from mouth to anus, some of them grow in particles of food or in what remains of food as it undergoes digestion. These microbes, speaking very strictly, are not parasites, or may not be; they merely take advantage of the

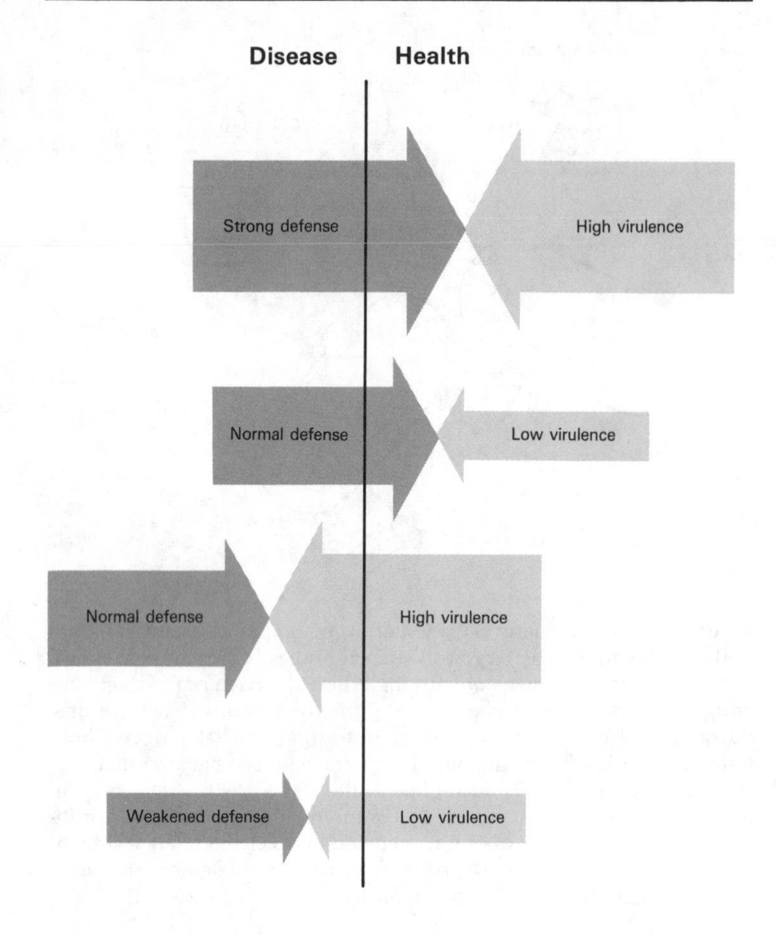

Figure 31 The delicate balance between disease and health.

warmth and moisture of the environment to live on our food or the products we digest it to, growing as they might grow on similar materials in an incubator. Some bacteria grow on the products produced by the activities of other bacteria.

After reading an article entitled "Life on the Human Skin" in January 1969, W. H. Auden was inspired to write the following elegant poetic contribution to the literature of microbial ecology entitled "A New Year Greeting."*

On this day tradition allots
  To taking stock of our lives,
My greetings to all of you, Yeasts,
  Bacteria, Viruses,
Aerobics and Anaerobics:
  A Very Happy New Year
To all for whom my ecoderm
  Is as Middle-Earth to me.

For creatures your size I offer
  A free choice of habitat,
So settle yourselves in the zone
  That suits you best, in the pools
Of my pores or the tropical
  Forests of armpit and crotch,
In the deserts of my forearms
  Or the cool woods of my scalp.

Build colonies: I will supply
  Adequate warmth and moisture,
The sebum and lipids you need,
  On condition you never
Do me annoy with your presence
  But behaving as good guests should,
Not rioting into acne
  Or athlete's foot or a boil.

Does my inner weather affect
  The surfaces where you live,
Do unpredictable changes
  Record my rocketing plunge
From fairs when the mind is in tift
  And relevant thoughts occur
To fouls when nothing will happen
  And no one calls and it rains?

I should like to think that I make

*From Auden, W. H., 1969, A New Year Greeting. *Scientific American,* 221 (6):134. Reprinted by permission of Curtis Brown, Ltd. Copyright © 1969 by W.H. Auden.

A not impossible world,
But an Eden it will not be;
My games, my purposive acts,
May become catastrophes there.
If you were religious folk,
How would your dramas justify
Unmerited suffering?

By what myths would your priests account
For the hurricanes that come
Twice every twenty-four hours
Each time I dress or undress,
When, clinging to keratin rafts,
Whole cities are swept away
To perish in space, or the Flood
That scalds to death when I bathe?

Then, sooner or later, will dawn
The Day of Apocalypse,
When my mantle suddenly turns
Too cold, too rancid for you,
Appetizing to predators
Of a fiercer sort, and I
Am stripped of excuse and nimbus,
A Past, subject to Judgment.

— *W.H. Auden*

GERMFREE ANIMALS

Microbiologists in the 1890s were well aware of the fact that humans and other animals are extensively inhabited by non-pathogenic microbes, and inevitably they began asking whether or not the microbes were necessary for health and continued life. Since an animal inside the womb of the mother is germfree, why not deliver a newborn animal under sterile conditions, feed it sterile food, provide sterile air, and see what happens? If all went well and the animal lived normally, known microbial species could then be deliberately introduced and subsequent changes, if any, could be observed. This was much easier said than done! There were obviously many technical problems to overcome. Nevertheless, the first experiments of this kind were attempted in 1895 by two German investigators. They delivered newborn guinea pigs by Caesarian section in a sterile chamber and rigged up devices for supplying sterile food and air. Many difficulties were encountered in keeping the animals alive; bacterial contamination was a frequent problem and eventually they gave up.

Other investigators then took up the challenge. From 1910 to 1915, Ernst Küster developed a better germfree apparatus (Figure 32). He experimented with goats and raised several germfree kids on sterile goat's and cow's milk for about 40 days. The experimental animal was maintained in a sterile box-like structure, and manipulations by the investigator were made by inserting hands and arms into rubber gloves (labeled "G") that extended into the box. Kids have a tendency to nibble at anything within reach, so to prevent them from chewing holes in the gloves the latter were rolled up after each use and tucked into metal-lidded recesses (see Figure 32*b*). Despite all precautions, it proved difficult to maintain sterile conditions for longer term experiments.

The next major developments occurred during the 1930s at the University of Notre Dame, where James Reyniers pioneered in designing elaborate germfree facilities that were dependable. He even built a large sterile room in which a person wearing a sterilized diving suit could tend to germfree animals (the researcher dons a diving suit and climbs into a tank full of disinfectant liquid before entering the germfree room through a hatch).

The technology for growing and studying germfree animals has become quite sophisticated, and research efforts over a number of decades now permit us to draw several important conclusions:

1. Animal life is possible without microbes, even for long periods, if proper diets are provided.

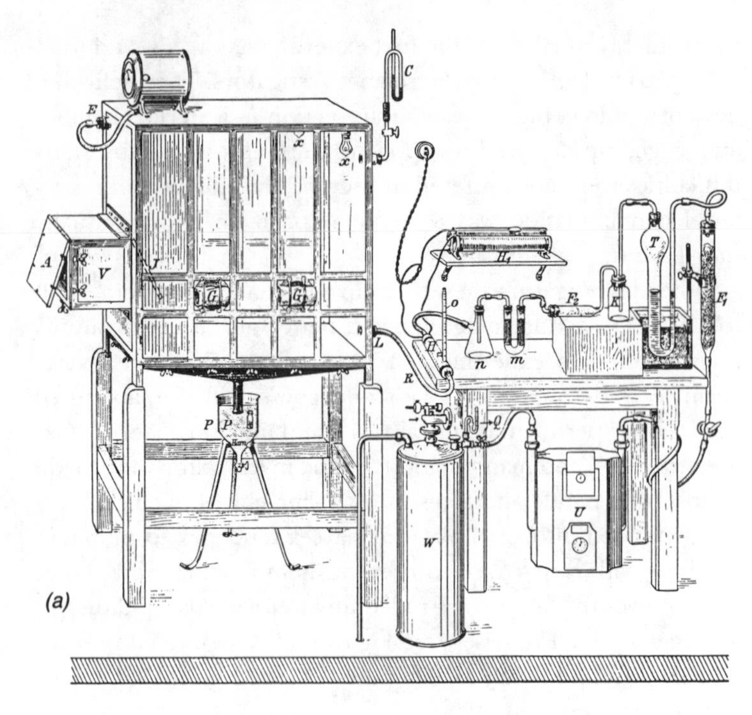

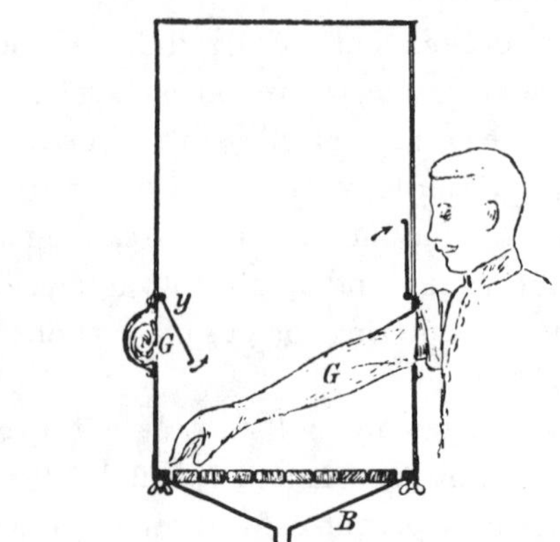

Figure 32 Küster's germfree apparatus. (a) The box in which the goat was kept is on the left. The air purification system is shown on the right. (b) A cross-section of the box.

2. The microbial population in the intestinal tract of certain animals produces vitamins that the animal uses for maintaining a healthy state.
3. In contrast to normal animals, natural body defenses against microbes in germfree animals are poorly developed.

These conclusions are supported by several lines of evidence. Germfree experimental animals become very ill, and may even die, if their diets are not supplemented with certain vitamins that do not have to be added to the diets of normal animals. This indicates that in normal animals, the microbial population in the intestine must furnish vitamins to the host. Additional direct evidence supporting this conclusion was obtained by analyzing fecal droppings from germfree chicks and normal chicks fed a diet that did not contain any vitamins of the B-complex. Droppings from the normal chicks were found to be rich in the vitamins; those from the germfree chicks were devoid of the vitamins, and these animals developed the symptoms of vitamin B deficiency diseases. In humans, it is thought that production of vitamins by intestinal microbes is probably not of importance to health, except when the diet is inadequate.

Children born with the condition known as "severe combined immune deficiency disease" soon die of microbial infections that do not respond to conventional medical treatments. When the mother of one such child became pregnant again, the parents agreed to a super-sterile Caesarian delivery. Within seconds of his birth in September 1971, the child, David, was placed in a plastic isolator, a chamber designed as a modification of germfree facilities developed at the University of Notre Dame. Extensive studies were made of David's immune system and other defense mechanisms as he grew and lived in a series of progressively larger germfree plastic "bubbles." These expanded gradually from a crib isolator to a four-chambered unit that included a playroom. He breathed sterile air, ate sterile food, and was outside of the bubble only once, in 1977, when he wore a sterile spacesuit designed and built for him by the National Aeronautics and Space Administration. David was kept free of infection throughout his entire life. He died at age 12 in 1984 from a massive proliferation of his own B-type white blood cells, which invaded a number of his internal organs. The abnormal growth of the B cells (a form of cancer) was possibly caused by a herpes-type virus.

The dire consequences of a total lack of natural defenses against microbes was a key element in one of the great classics of early science fiction. A radio broadcast in 1938 based on this classic created considerable public alarm; see Appendix III.

19

Viruses Confound Microbe Hunters

Are viruses living organisms or not? They have a number of properties attributable to living cells, but lack others. Viruses are incapable of multiplying by themselves. They can multiply only inside susceptible living cells (microbial, plant, or animal) and in many respects are perfect parasites. Pioneering studies on viruses that attack bacteria laid the groundwork for analyzing details of the mechanisms employed by viruses that invade and multiply in plant and animal cells. We now have detailed knowledge of the structures of many viruses and of their replication processes. However, aside from the use of vaccination for humans and animals to prevent some diseases, we still do not have effective means of aborting virus growth in plant and animal tissues.

The first virus was discovered in 1898 as the causative agent of tobacco mosaic disease, also known as "leaf spot." The infective agent could be transmitted from sick plants to healthy plants by sap that contained *no* microbes at all. Careful examination of the sap with the best optical microscopes revealed no structures that could be identified with the infectious agent. Clearly, the culprit was much smaller than typical bacterial cells. As the years passed, a number of other diseases of plants and animals could be shown to be due to "invisible" viruses, for example: smallpox, yellow fever, poliomyelitis, and measles.

The true nature of viruses remained obscure for a number of decades, and the mystery was compounded in 1935 when W. M. Stanley (1904–1971), an American scientist, reported that tobacco mosaic virus could be crystallized. It then seemed that viruses surely could *not* be any kind of living cells; cells cannot be crystallized. Could they be some kind of inanimate, but complicated molecules? There was a dilemma: whatever they were, viruses multiplied with the dynamic qualities of life. It appeared that viruses were in limbo, somewhere between "life" and "nonlife."

In the mid-1930s, a new kind of more powerful microscope was developed, the electron microscope, which uses a beam of electrons instead of a beam of light to view the subject. The electron beam is focused on the specimen (in a vacuum) with magnets, in contrast to the glass lenses used in the optical microscope. The electron microscope provides much better resolution of very small objects. After the end of World War II, microbiologists were finally beginning to see what viruses actually looked like, and they were indeed interesting: geometrical structures, of different degrees of complexity, approximately 10 to 100 times smaller than bacteria (Figure 33).

The great breakthroughs in our understanding of viruses came from study of *bacteriophages*, viruses that attack bacteria (Figure 34). The scientific "truth" in the following verse now becomes clear:

Big fleas have little fleas
Upon their back to bite 'em
The little ones have lesser ones
And so *ad infinitum*

Bacteria grow much more rapidly than animal and plant cells, and it is consequently much simpler to do many kinds of ex-

Bacteria

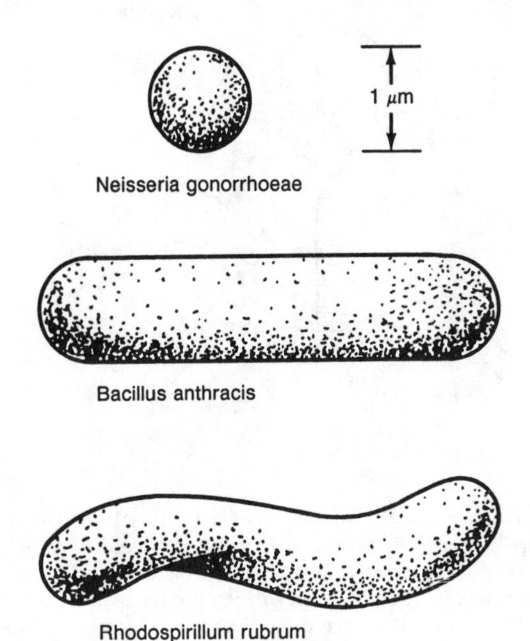

Viruses		**Proteins**	
Influenza	o	Hemoglobin	·
Polio	•	Egg albumin	·
Bacteriophage		Antibody	·

Figure 33 Relative sizes (approximate) of some bacteria, viruses, and protein molecules. Reference diameters: *Neisseria gonorrhoeae* cell, 1 μm; influenza virus, 0.1 μm. The hemoglobin molecule measures 0.003 × 0.015 μm. One μm (micrometer) equals one-millionth of a meter.

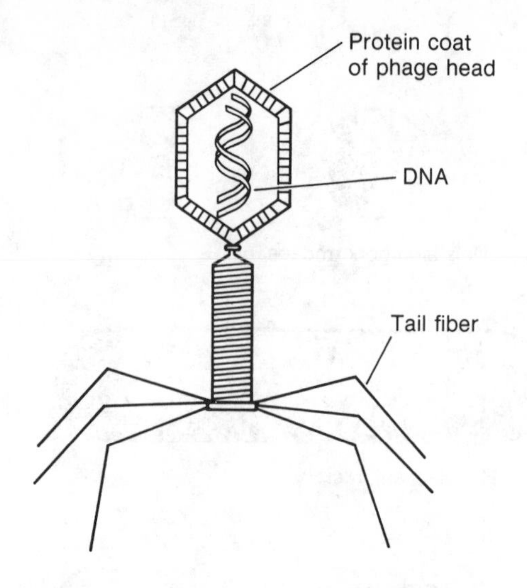

Figure 34 Diagram of the structure of a bacterial virus (bacteriophage; "phage"). Reproduction of the virus is initiated by attachment of its tail fibers to the surface of a susceptible bacterial cell. A syringe-like action then injects the phage DNA into the bacterium. Metabolic systems of the parasitized cell are then diverted to biosynthesis of new phage heads, tails, etc. (instead of to synthesis of normal cell components). After about 30 minutes, the cell bursts liberating 100 or more new phage particles.

periments with bacteria and with the viruses that attack them. The fundamental aspects of virus multiplication turned out to be essentially the same for bacteria, plant, and animal viruses. There are many differences in details that are important for understanding how to combat infectious plant and animal viruses, but the basic features of virus multiplication were first revealed most clearly by studying bacteriophages.

The features that distinguish viruses from all other living organisms are as follows:

1. Viruses lack the extensive biochemical enzyme machinery required to reproduce themselves. They contain only a few highly specialized enzymes needed for certain steps

of virus fabrication, but they are devoid of other enzyme systems, for example, for generating the energy (ATP) necessary for synthesis of virus particles.

2. Viruses resemble living cells in that they carry the genetic "instructions" (in the form of nucleic acid) for new virus synthesis, but viruses cannot grow independently in ordinary nutrient media the way microbes can. Viruses can multiply *only* inside of susceptible living cells.
3. When a virus particle invades a host cell, it "takes over" the biochemical apparatus of the cell and "regears" it for the production of new virus particles. This subversion destroys the host cell, but the virus thrives. In the case of the bacteriophage and bacterium system, a single "phage" particle attacks a single bacterium, and after about 30 minutes, the bacterial cell burst, liberating about 200 new "phage" particles.
4. Viruses contain a single kind of nucleic acid, either DNA or RNA, which is packaged within a protein "coat"; in contrast, cells always contain both DNA and RNA.

In summary, viruses lack the biochemical mechanisms needed for their own multiplication. In this sense, they could be called "incomplete organisms." Viruses are inert and can be stored for long periods without loss of infectivity. They acquire the ability to reproduce themselves only when inside suitable living cells.

Since viruses differ in significant ways from microbes (as well as other cells), why is the study of viruses always considered to be part of microbiology? First, the very small sizes of viruses immediately places them in the domain of biological entities that includes microbes. Also, if we were to consider the properties of all the known kinds of *bacterial* parasites of bacteria, we would find that the borderline between viruses and tiny parasitic bacteria becomes quite indistinct. Despite their idiosyncracies, viruses resemble microbes in a number of respects, and interact with higher organisms in the same fundamental ways that pathogenic microbes do. The question of whether or not viruses should be considered to be "living" has been responsible for endless philosophical discussions of academic interest. Nonetheless, for the reasons stated above, there is general agreement that viruses are more closely connected to the microbial universe than to other categories of living things.

20

The Control of Microbial Disease

A variety of procedures are used to kill or inhibit the growth of potentially pathogenic microbes in our food and drink, on objects that we contact daily, and on or in the body. They vary greatly in selectivity—some kill microbes indiscriminately, whereas others may affect only a few species of closely related organisms. The antimicrobial weapons now available are of two general kinds: physical agents and chemical agents.

PHYSICAL AGENTS

Heat is the most common physical agent used for killing microbes in food and on objects (sterilization). Different heating regimens are employed depending on the circumstances. For example, sterilization of canned foods requires relatively high temperatures and a long duration of heating to ensure destruction of spores of pathogens. In contrast, the pasteurization process involves mild heating of a short duration. Pasteurization is aimed at killing the relatively heat-sensitive pathogenic bacteria likely to contaminate milk and other liquids; not all the microbes present are killed. Various types of radiation can be used to sterilize the surfaces of objects, but in practice the most widely used is ultraviolet radiation. This type kills cells by affecting the structure of DNA, whereas killing by heat is due to adverse effects on proteins.

CHEMICAL AGENTS

Numerous classes of chemical substances can inhibit the growth rates of microbes or can kill them. Certain effective chemical agents can be used externally, on the skin or to disinfect objects. Others, such as antimetabolites and antibiotics, are used internally in order to fight disease. These chemicals are very important in *chemotherapy*, the treatment of an infectious disease using drugs.

Antiseptics These are chemicals that can be safely applied to the skin or mucous membranes; for example, 70 percent alcohol, tincture of iodine (iodine in alcohol), and phenol.

Disinfectants Chemical agents used to kill microbes in or on inanimate objects or materials. Chlorine gas and sodium hypochlorite, a compound of chlorine, are good examples. These are widely used for disinfection of water supplies and dairy equipment, and of eating utensils in restaurants. Clorox,* a well-known household bleach, is simply a 5 percent solution of sodium hypochlorite; one tablespoon in a gallon of water makes a good disinfectant preparation.

Antimetabolites A *metabolite* is a chemical substance that plays some part in metabolism. Needless to say, there are a large

---

*Clorox ® is a registered trademark of The Clorox Co., Oakland, California.

number of metabolites involved in the normal growth of microbes. An *antimetabolite* is a manmade chemical that is very similar in structure to an essential normal metabolite, but different enough to obstruct processing of the normal metabolite. Thus, the microbe is deceived into using the antimetabolite compound, and this usually results in formation of abnormal proteins or vitamins that are nonfunctional. Antimetabolites ordinarily do not kill microbes, but they can slow down their growth rates drastically. This can give the body's natural defense mechanisms advantages in dealing with the invader. The so-called sulfa drugs, widely used at one time for treating infectious diseases, are examples of antimetabolites.

ANTIBIOTICS

Antibiotics are defined as chemical substances produced by certain microbes that are able, at low concentrations, to kill other microbes or inhibit their growth. The existence of antibiotics, the most useful and potent agents for fighting infectious disease, was discovered by accident in 1929 by Alexander Fleming, a microbiologist working at St. Mary's Hospital and Medical School in London. He was hired by Almroth Wright, a famous physician, who devoted much of his time to research on how immunity to typhoid fever could be achieved. Wright was a close friend of George Bernard Shaw, who frequently visited Wright's laboratory in the evening, after the theater, for a cup of tea. Wright often did his research from evening until 3 or 4 A.M., and Shaw made him the hero of his play "The Doctor's Dilemma." (The play contains quite a lot of microbiology, including talk about white blood cells eating microbes.)

Fleming's research centered on ways of killing pathogenic bacteria with antiseptics, and he frequently used staphylococci as the test organism. Fleming was not a particularly tidy researcher; in fact, he was often teased for being disorderly. Typically, his laboratory bench was piled high with old Petri dish

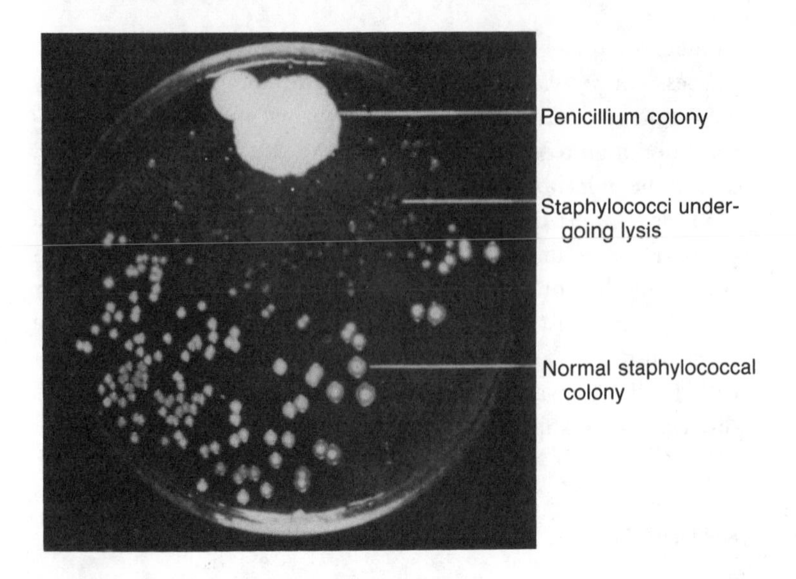

Figure 35 Fleming's photograph of the original Petri dish on which he observed "dissolution of staphylococcal colonies in the neighbourhood of a penicillium colony."

cultures that should have been discarded. One day in 1929, while talking to a young assistant, he lifted the lids of a few old dishes and glanced at the agar cultures. These had become contaminated with molds; this frequently occurs when cultures are allowed to sit around for months. He muttered to his assistant: "As soon as you uncover a culture dish something tiresome is sure to happen. Things fall out of the air." Suddenly, he stopped talking, and then said, "That's funny. . . ." He was struck by an unusual sight. On the particular dish he was examining, there was a large fungus colony on the agar next to where he had been growing some yellow colonies of *Staphylococcus* bacteria. However, the bacterial colonies near the fungus growth on this dish seemed to have dissolved and looked like small drops of dew (Figure 35). Fleming eventually identified the fungus as *Penicillium notatum*, which naturally secretes an organic chemical substance of relatively simple structure that kills a number of bacterial species very effectively. The substance was appropriately named *penicillin*, and it became the first antibiotic to be discovered.

It is clear that Fleming did not realize the potential value of penicillin for treatment of infectious diseases, and it was not until 1938 that this idea began to take root. Early in that year, Ernst Chain of the University of Oxford came across Fleming's 1929 report and convinced his department chairman, Howard Florey, that further research on penicillin would be of interest and of scientific value. In an interview given in 1967, Florey said:

> There are a lot of misconceptions about medical research. People sometimes think that I and the others worked on penicillin because we were interested in suffering humanity—I don't think it ever crossed our minds about suffering humanity; this was an interesting scientific exercise. Because it was some use in medicine was very gratifying, but this was not the reason that we started working on it. It might have been in the background of our minds; it's always in the background in people working in medical subjects. . .but that's not the mainspring.

In any event, by 1940, Chain and Florey and their colleagues were in the midst of a rapidly expanding pioneering effort to isolate penicillin in pure form and test its chemotherapeutic effects on bacterial infections of humans. The first "miraculous" cures were effected in 1941 and led inevitably to a burst of research activity aimed at finding other antibiotics. In 1954, Florey, Chain, and Fleming were awarded a Nobel Prize for their pioneering work; since then over 1000 antibiotics from various fungi and bacteria have been isolated and characterized.

Different classes of antibiotics have different mechanisms of action. For example, penicillin interferes very specifically with synthesis of microbial cell walls; as a result, the microbe bursts and dies. Streptomycin inhibits protein synthesis in susceptible bacteria. Other antibiotics cause disruption of bacterial cell membranes.

It is worth noting that the path from basic discovery to practical application is frequently a long, hard road. The story of penicillin is an excellent case in point. The original strain of *Penicillium notatum* studied by Fleming produced relatively small amounts of penicillin. A related organism, *P. chrysogenum*, isolated in 1951, was more useful; it produced about 60 milligrams (mg) of the antibiotic per liter of growth medium. However, this was still too small a yield to form the basis of an industrial isolation process. Over a number of years, several groups of scientists systematically investigated *P. chrysogenum* with the aim of isolating mutant strains that secreted more of the antibiotic. Strain E-15.1, the "final strain," produces 7000 mg of penicillin per liter, and after other improvements, the yield is now up to about 20,000 mg per liter.

Are there any antibiotics active against viruses? Those effective against microbes are known to be ineffective for treatment of virus diseases of animals. Since viruses multiply by exploiting the biochemical machinery of host cells, it is understandable that it will be difficult to find chemicals that inhibit virus growth without also adversely affecting cells of the host. More detailed knowledge of the biochemistry and molecular biology of viruses, however, can be expected to eventually enable us to find drugs that will do the trick.

21

The Role of DNA and New Vistas in Microbial Biotechnology

After World War II ended, research in microbiology and biochemistry increased rapidly. Many basic problems were seen in sharper focus and various new techniques were developed, especially methods for separating and characterizing important cell components. A "golden age" of biochemistry began in the early 1950s, leading to extensive exploration and mapping of the workings of growing cells. This period also marked the beginnings of sophisticated insights into mechanisms of bacterial reproduction. Discoveries made starting in about 1948 showed for the first time that bacterial cells have the capacity to conjugate and exchange genes. It was previously thought that bacteria were primitive in all respects and multiplied only by nonsexual means, namely, by simple division of a cell into two daughter cells after the mother cell had grown to some critical size. With the demonstration of a sexual process in bacteria, the field of bacterial genetics was born.

At about the same time, research with bacteria that cause pneumonia showed that genes are made of DNA. These discoveries started an avalanche of research activities that gradually became known as molecular biology. One prominent feature of research in this field is the deliberate experimental transfer of particular genes from one bacterium to another, or from a bacterium to cells of a higher organism (or vice versa). Genetic engineering of this kind is largely based on use of bacteria and their viruses, and it is being eagerly exploited in fundamental research on biological mechanisms and in commercial biotechnology. In this chapter we will see how DNA performs its important role and how it can be used for genetic engineering.

The science of genetics deals with the mechanisms by which the hereditary properties of cells and organisms are determined and transmitted from one generation to the next. Early research indicated that this continuity must be governed by "factors," that were later called *genes*, long before there was any idea of their mechanism or chemical composition. Despite this lack of information, geneticists showed that genes of animal and plant cells are arranged linearly in microscopically visible "bodies," the chromosomes. When we say that reproduction in higher organisms occurs by sexual recombination, we mean that genes in chromosomes from both parents contribute to the genetic makeup of the offspring. In all eucaryotes, including microbial eucaryotes, the fundamental principles of sexual recombination are basically similar. Despite numerous variations in details, there are always two mating types, male and female, or equivalents designated as plus and minus.

Before 1946, there was no evidence for the occurrence of mating types or sexual recombination in bacteria; in fact, many scientists thought that procaryotes simply did not have processes of this kind. At that time, however, ingenious experiments revealed that bacteria do indeed have ways of exchanging genes. Ironically, it turned out that the study of genes and gene exchange in bacteria (and in bacteriophages) provided the means of deciphering the mechanisms by which genes control the properties of *all* organisms. Eventually, it was established that genes consist of DNA (deoxyribonucleic acid). Thus, some understanding of DNA structure is essential for even an elementary appreciation of the exquisite mechanisms involved in gene action.

STRUCTURE OF DNA

DNA is a large macromolecule composed of three kinds of chemical units that are arranged in a very specific manner. Two of these units provide the backbone of DNA in the form of a two-stranded helix, in which two coiled fibers are connected (Figure 36). The backbones are quite monotonous in that they consist simply of alternating units of phosphate and of a five-carbon atom sugar called *deoxyribose.* If phosphate is designated by a black circle, and the sugar as "D," a single backbone could be represented as:

••• ---●---D---●---D---●---D---●---D--- •••

The two backbones of DNA are held together by pairs of *nucleic acid bases.* These bases represent the third kind of unit in DNA and consist of four types of small nitrogen-containing molecules that have distinctive chemical properties.

Base	Designation
Adenine	A
Thymine	T
Guanosine	G
Cytosine	C

Each base is connected to a unit of D (deoxyribose) in the backbone, as shown in Figure 37. Note that the bases pair up

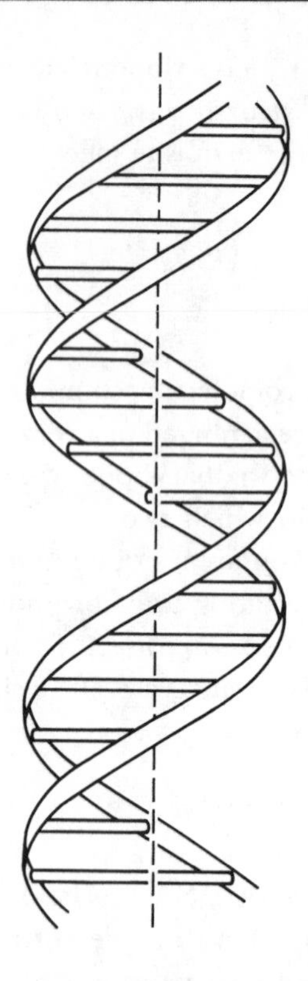

Figure 36 Schematic representation of the DNA double helix. The helical strands are composed of alternating units of a five-carbon sugar (deoxyribose) and phosphate. The "rungs" represent chemical bonds between pairs of nitrogen-containing nucleic acid bases that extend toward the axis of the helix (which is indicated by a dashed line).

in a special way. There are only two kinds of pairs: A· · ·T and G· · ·C (the dots here represent a type of chemical bond). In DNA the number of adenine molecules always equals the number of thymine molecules. Likewise for guanosine and cytosine.

How can a structure of this kind possibly contain the large amount of information needed for construction of a new cell? The magic of DNA is that it consists of a unique *coding* system that can specify the structures of a large number of different proteins. The number of proteins in a typical bacterial cell is

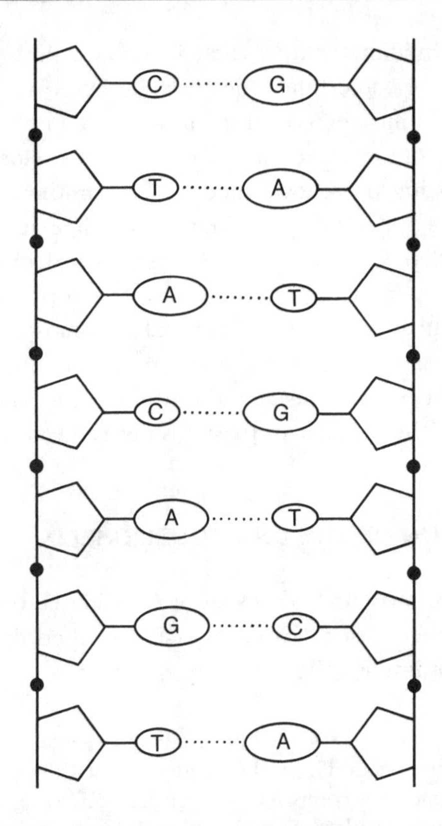

Figure 37 Diagram of a portion of a DNA molecule (uncoiled and flattened). The molecule has a ladder-like structure, with the two uprights composed of alternating sugar (the pentagons represent deoxyribose sugar units) and phosphate groups (•). The "cross rungs" are composed of pairs of nucleic acid bases, each pair linked by a chemical bond.

estimated to be about 3000 (no one knows for sure), and each is coded by a *gene*. Each gene contains about 1000 bases, and the coding capacity of the DNA in a cell is more than sufficient to account for all the proteins that have to be made.

In contrast to higher organisms, bacteria have only a single *chromosome* per cell. It consists of a single long molecule of DNA. Stretched out to its full length, the DNA of a single cell would be about 1 millimeter long, that is, about 1000 times longer than the entire bacterium! Obviously, in the living cell the DNA chromosome must be coiled up in a very compact form. The bacterial chromosome exists in the cell as a closed circle of DNA consisting of genes joined to one another. In addition, many species of bacteria contain, in each cell, a number of rings of DNA called *plasmids*. These range in size from tiny circles of DNA, containing only a few genes, to enormous structures that are almost as large as bacterial chromosomes, carrying hundreds of genes. When bacteria grow and multiply, the chromosome and plasmids usually duplicate at the same rate; this ensures that the number of plasmids per cell remains constant.

REPLICATION OF THE DNA DOUBLE HELIX

Details of the structure of DNA were elucidated by James Watson and Francis Crick in 1953. Crick described the essence of DNA replication as follows:*

> The genetic message is conveyed by the exact base-sequence along one chain. Given this sequence, then the sequence of its complementary companion can be read off, using the base-pairing rules (A with T, G with C). The genetic information is recorded twice, once on each chain. This can be useful if one chain is damaged, since it can be repaired using the information—the base sequence—of the other chain. . . .Because they fit together so precisely, each chain can be regarded as the mold for the other one. Conceptually the basic replication mechanism is very straightforward. The two chains are separated. Each chain then acts as a template for the assembly of a new companion strain, using as raw materials a supply of four standard components. When this operation has been completed we shall have two pairs of chains instead of one, and since to do a neat job the assembly must obey the base-pairing rules (A with T, G with C), the base-

*From Crick, F., 1981, Life itself; its origin and nature. Simon and Schuster, New York.

sequences will have been copied exactly. We shall end up with two double helices where we only had one before. Each daughter double helix will consist of one old chain and one newly synthesized chain fitting closely together, and more important, the base-sequence of these two daughters will be identical to that of the original parental DNA. The basic idea could hardly be simpler.

TRANSLATION OF DNA CODE TO PROTEIN STRUCTURE

Most cell proteins act as enzyme catalysts; some have structural functions, acting as bricks and mortar would. Each kind of cell protein is different in composition in that its sequence of amino acids is different. Thus, we encounter the fundamental question: how does a gene specify the exact sequence of amino acids that are assembled to fabricate a particular protein?

Since there are 20 kinds of amino acids, the essence of gene action is translation of the "four-letter DNA language" (A, T, G, C) into the "20-word protein language." Research over the past several decades has revealed how this is accomplished. The mechanism is extremely complicated, and involves many cell components. Aside from the complexity, the process is extraordinary in several other respects. Cell proteins have to be synthesized

- economically,
- in the right quantities, and
- with high fidelity, that is, the amino acids must be sequentially attached in the correct order.

Clearly, many controls must be built into what can only be described as an exquisite production device. We will consider the mechanism only in bare outline.

The double-stranded helix separates, yielding two strands, one of which is used as a template or "blueprint." To simplify representation of the blueprint code, we will use a notation that shows only the sequence of bases in the template strand. At this point, we are not concerned with details of DNA chemistry, but wish to focus only on the sequence of bases as they occur in the DNA strand. A hypothetical sequence could be

—A—T—T—G—C—A—T—C—G—T—G—G—

Further progress in understanding the mechanism requires the knowledge that a sequence of three particular bases specifies insertion of a particular amino acid into the fabric of a protein molecule that is being assembled. Below, we see how this rule translates the hypothetical sequence and dictates insertion of four amino acids (designated as Ile, Ala, Ser, and Trp) in correct order into a "growing" (still incomplete) protein molecule.

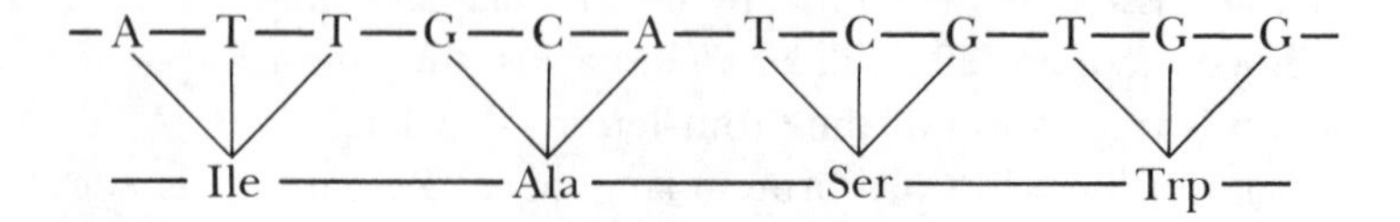

Thus, –A–T–T– specifies amino acid Ile, –G–C–A– specifies amino acid Ala, and so on.

The size of an average protein molecule is about 300 to 350 amino acid units. If it takes three nucleic acid bases to specify insertion of each amino acid unit, it follows that an average gene consists of 900 to 1000 bases lined up in a unique sequence. Each protein begins with a particular amino acid, and ends with a particular amino acid. The DNA must therefore provide code words for "start" and "stop"; these also consist of special sequences of three bases.

An important aspect of the genetic code is that with rare exception, the code for conversion of DNA language to protein language is the same throughout the biological world, from viruses to humans. This strongly reinforces the idea that all living things descended from a common ancestor. One of the numerous major problems in trying to understand the origin of life is the question of how the genetic code evolved from some simpler version to the one that exists today.

GENE MUTATION

Since genetic information is present in DNA in the form of particular sequences of bases, this information can also be thought of as messages written in a four-letter language (the four letters being A, T, G, and C). Using this analogy, a mutation, then, would be equivalent to a misprint in a line of type. A single mutation would occur when a single letter (single base) had been deleted or incorrectly inserted in the message, or if an incorrect letter has been substituted for the correct one. Using a song title analogy, we can illustrate how the meaning of a genetic message can change when a single letter "misprint" (mutation) occurs:*

"Rock around the *c*lock"
becomes
"Rock around the *b*lock"

In a microbial cell, a single base change or alteration of the correct base sequence (by deleting or inserting a base) has the usual consequence that the particular protein specified by the gene in question will either not be formed or will be produced in an abnormal nonfunctional form. If this protein happened to be an enzyme required for the production of the amino acid methionine, the bacterium would now be a "methionine mutant," that is, a strain unable to grow unless preformed methionine is supplied in the medium. A large variety of microbial mutants have been isolated, each with defects in the biosynthesis of amino acids, vitamins, or other cell components.

---

*From Rosenberg, E., and Cohen, I. R., 1983, *Microbial Biology*. Saunders College Publishing, New York.

Mutants occur spontaneously in microbial populations due to random events that affect proper sequencing of bases during DNA synthesis. Such changes are rare—the average gene may be duplicated one million times before a single detectable mutation occurs. The frequency of mutation, however, can be greatly increased by exposure of cells to radiation, such as X rays or ultraviolet light, and to a large assortment of so-called mutagenic chemicals. Nitrite, which is used as an additive to certain foods, is an example of such a chemical.*

In summary, changes in the normal sequence of bases in DNA result in "misreading" of the genetic code. This usually leads to formation of proteins that do not function properly, either in their roles as catalysts (enzymes) of metabolism or as parts of the structural framework of the cell. For a more detailed (but easy to understand) description of how mutations result in altered proteins (that is, altered reading of the genetic code), see the excellent article by Francis Crick entitled "The Genetic Code" in Scientific American, October 1962.

GENE TRANSFER (GENETIC RECOMBINATION) IN PROCARYOTES

The first demonstration of gene transfer (genetic recombination) in bacteria was made possible by exploiting mutants of *Escherichia coli*. The classic experiment is succinctly summarized as follows:†

> Mutant strains deficient for two or more growth factors were produced by irradiation of a strain of *Escherichia coli*. Two strains,

---

*Some mutagenic agents "spoil" DNA language by changing a nucleic acid base to a modified form that will not engage in the proper kinds of base-pairing in the DNA double helix. Thus, nitrite can modify base A to a form that pairs with C instead of to T.

†From Luria, S. E., 1947, Recent advances in bacterial genetics. *Bacteriological Reviews, 11*, p. 1–40. The term *minimal medium* used in this quotation refers to medium not containing the growth factors.

> each carrying a different pair or group of biochemical deficiencies (double biochemical mutants), were then grown together in a complete liquid medium. After growth, large inocula were plated on minimal medium agar on which neither of the two strains could grow. Colonies appeared, consisting of cells that had permanently acquired ability to grow on the minimal medium like the original strain of *Escherichia coli*. These cells must therefore have the ability to synthesize all four growth factors, combining the synthetic powers of the two parent strains.

This kind of genetic recombination occurs by a mechanism called *conjugation*. In a population of *E. coli* cells, certain of them act as males (called F^+) and others as females (F^-). Figure 38 illustrates conjugation between F^+ cells of one kind of mutant and F^- cells of another kind. First, a "conjugation bridge" is formed between an F^+ cell and an F^-. The circular F^+ chromosome breaks at a certain point and begins to thread into the F^- recipient. Genes on the F^+ chromosome (indicated by A through E) are transferred into the F^- cell in the order that they were aligned in the donor (male) bacterium. Ordinarily, the entire chromosome is not transferred because the conjugation bridge is very fragile and ruptures spontaneously. The transferred portion of the F^+ chromosome, or part of it, is incorporated into the recipient F^- chromosome, yielding a bacterium with a new assortment of genes.

Assume that in the foregoing example, the recipient cell had two damaged genes that made it a double biochemical mutant, requiring provision of two growth factors for multiplication. If the piece of donor F^+ chromosome that was incorporated contained the corresponding *undamaged* genes, the *recombinant* (recipient) bacterium would now have the ability to make the growth factors in question and would not need them supplied in the medium. This kind of experiment was the means by which the existence of gene exchange in bacteria was first established.

In addition to conjugation, there are other natural mechanisms for transfer of genes between bacterial cells. For example, certain bacteriophages (viruses) are able to transfer chromo-

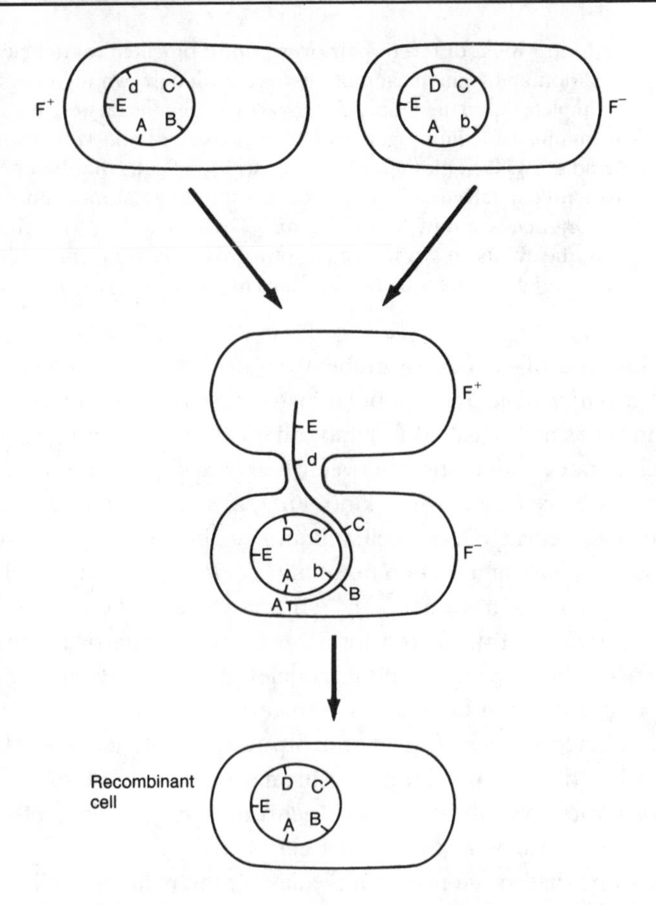

Figure 38 Conjugation between F^+ and F^-. In this representation, only a few genes (A, B, C, D, and E) of the bacterial chromosome are indicated. The F^+ cell contains normal DNA, except for one damaged gene designated as d. The F^- cell contains one damaged gene indicated as b. In the center panel the chromosome from an F^+ cell is seen threading into an F^- cell through a conjugation bridge. After the chromosome fragment carrying genes A, B, and C had entered the F^- cell the cells were separated by agitating the suspension in a blender. A recombination event then occurs resulting in replacement of damaged gene b by normal gene B. The recipient F^- cell is now a "genetic recombinant." If gene B happened to be a gene controlling production of a particular vitamin, the recombinant cell would have the ability to make the vitamin; the original F^- cell with the b gene was unable to produce the vitamin.

somal genes from one bacterium to another. Detailed study of such virus-mediated gene transfers and of the genetics of *E. coli* and other procaryotes has provided the basic framework of the body of knowledge now referred to as molecular genetics.

GENETIC ENGINEERING OF MICROBES

Microbes usually make metabolic products only in amounts required for survival and reproduction. One aim of genetic engineering of microbes is to deliberately alter genetic composition so as to create strains that produce excessive quantities of useful substances. In 1973, new techniques were developed that made it possible to transfer genes from almost any source into various bacteria, including *Escherichia coli,* or into yeast cells. Plasmids are the favorite vectors for such procedures. A typical procedure involves these steps:

1. a particular gene is isolated from, for example, a human or plant cell;
2. the gene is spliced into an *E. coli* plasmid ring;
3. the plasmid is reintroduced into *E. coli* cells; and
4. the recombinant cells are grown in large quantity with the plasmids replicating many times during growth.

The culture ends up containing billions of copies of the "cloned" human or plant gene. Steps 1 and 2 were made possible by the discovery of DNA "restriction enzymes." These remarkable enzymes (obtained from bacteria) are able to cut double-stranded DNA at specific sites; there are many kinds of restriction enzymes, each of which cuts DNA at different base sequences. The basic features of recombinant DNA methodology are summarized in Figure 39.

What are the purposes of these procedures? First, they are powerful tools for basic research on unsolved problems such

Figure 39

THE TECHNIQUE IS CALLED

GENE CLONING,

AND IT WORKS LIKE THIS:

FIRST, CHOOSE A HUMAN GENE ENCODING SOME USEFUL PROTEIN.

FOR YOUR BACTERIAL DNA, YOU NEED SOMETHING THAT WILL BE REPLICATED ONCE IT'S RETURNED TO THE CELL — A "***VECTOR***", SO-CALLED.

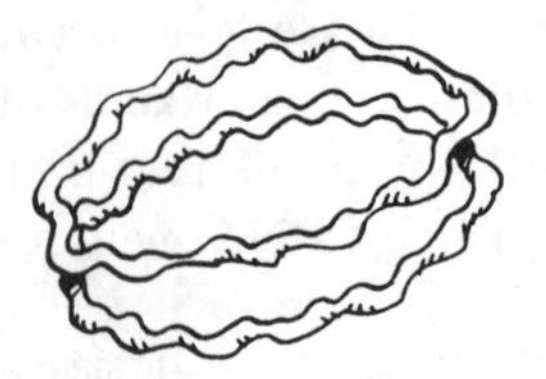

LUCKILY, *E. COLI* HAS SMALL RINGS OF DNA CALLED ***PLASMIDS***, SEPARATE FROM THE CHROMOSOME. YOU CHOOSE (OR ENGINEER!) A PLASMID CONTAINING THE SEQUENCE *G·A·A·T·T·C*, AND REMOVE IT FROM THE BACTERIUM.

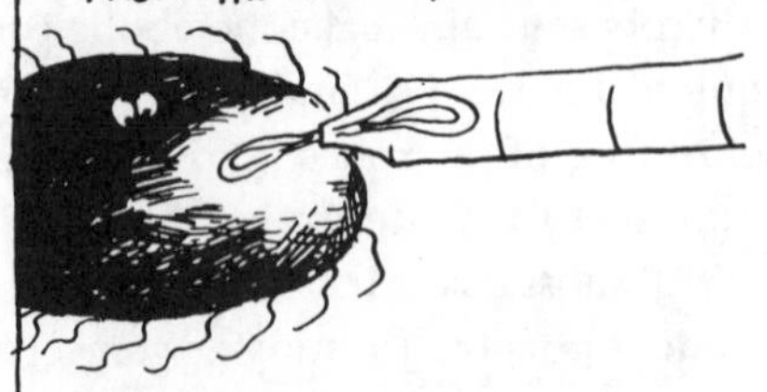

JUST AS ABOVE, YOU ***SPLICE*** THE HUMAN GENE INTO THE PLASMID—

AND PUT IT BACK INTO ***E. COLI***.

BE FRUITFUL AND DIVIDE!
NOW YOU FEED THE BACTERIUM AND LET IT BREED.
THE PLASMID IS REPLICATED ALONG WITH EVERYTHING ELSE IN THE BACTERIUM.
WITHIN A FEW HOURS, WE CAN HAVE A BILLION BACTERIA IN A FEW DROPS OF CULTURE— AND A BILLION COPIES OF THE HUMAN GENE!!
IF WE'VE INCLUDED THE PROPER REGULATORY REGIONS AS WELL, THE BACTERIA SHOULD EXPRESS THE GENE, AND WE CAN EXTRACT SUBSTANTIAL AMOUNTS OF THE HUMAN PROTEIN. MIRACULOUS!
PRAISE BE!
INSULIN
INTERFER

as the mechanisms of developmental biology. Salvador Luria describes a relevant basic research application as follows:*

> Imagine that a scientist wants to introduce into a bacterium a fragment of human DNA presumed to contain a cancer gene. He will use one or more of the restriction enzymes to cut the DNA from human cancer cells into fragments and also cut an appropriate bacterial DNA into similar pieces. Then he mixes the sets of fragments and adds to the mixture a sealing enzyme that rejoins the pieces together. Some piece containing the cancer gene will join up with the bacterial DNA. Then the bacterial DNA can be made to enter intact bacteria, where the cancer gene can become part of the bacterial gene string and can be further isolated and identified.

Second, there are almost unlimited possibilities for practical applications in medicine, agriculture, and other spheres of interest to modern society. The most familiar example comes from the pharmaceutical industry. If a human gene spliced into an *E. coli* plasmid functions properly, the *E. coli* host cells will produce the protein that the human gene specifies *in substantial quantities*. This can now be done with the human insulin gene. Thus, the commercial process for obtaining the valuable protein insulin, needed by diabetics, can be radically altered. Instead of grinding up trainloads of pig pancreas glands and isolating the pig insulin from a great mixture of other kinds of pancreas proteins, we can engineer *E. coli* cells to make human insulin for us. In principle, it is now feasible to make almost any protein in a similar fashion.

REMARKS ON THE HISTORY OF BIOTECHNOLOGY

Biotechnology is a commonly used term in today's mass media, and it has taken on many different meanings. One interesting

*From Luria, S. E., 1984, *A Slot Machine, a Broken Test Tube*. Harper & Row, New York.

definition of the aim of biotechnology is to bring engineering, biochemical, and microbiological techniques together as elements of novel art, as distinct from refinements of ancient art. A driving force in this important field is the anticipation of producing commercially valuable results and the hope that new remedies for effectively alleviating human suffering are forthcoming.

Microbes loom large in any discussion of modern biotechnology; indeed, they were the predominant biological systems exploited in earlier times.* The unconscious domestication of microbes by ancient civilizations for producing alcoholic beverages has already been discussed. Other microbe-catalyzed processes have been used for centuries, such as preservation of foods by natural acidification, now known to be due to formation of organic acids from sugars by anaerobic bacteria. Sauerkraut is a good example. Ancient recipes worked well; cabbage was cut into small pieces, salt was added to make juice and sugar come out of the cabbage cells, and the whole mass was loaded down with planks and stones. This last step helped to impede access of air, establishing anaerobic conditions; sauerkraut thus became the earliest known form of silage. In 1739 a Hungarian army doctor recommended that sailors should consume sauerkraut every day to avoid contracting scurvy (cabbage has a high content of vitamin C).

During World War I, the industrial-scale use of microbes for fermentative production of acetone and other organic compounds became a high priority of several governments. Acetone was used in the manufacture of explosives and the lacquers employed for coating the canvas wings of airplanes and was in critically short supply. Distilleries in Canada and the United States were converted to facilities for microbial fermentation of carbohydrates (in maize) to a mixture of acetone and alcohol. Subsequently, still other possibilities for harnessing the chemical activities of microbes for useful purposes became apparent.

---

*A remarkable account of the benefits of do-it-yourself microbial biotechnology under difficult circumstances (in a prisoner-of-war camp) is given in Appendix IV.

Between 1935 and 1955, research in cell biochemistry escalated dramatically, and the major features of enzyme action and metabolic patterns were elucidated. This was facilitated in part by the introduction of new, more sensitive and powerful techniques for analyzing the dynamic biochemical processes of bacteria and other types of cells. Of special note in this connection was the use of radioactive tracers to follow the fates of carbon and other atoms in their complex transformations during metabolism. The same period was also noteworthy in other ways. The great potential of natural antibiotics for treating microbial infections was fully realized for the first time. A remarkable effort by teams of scientists from several countries culminated in the industrial-scale isolation of penicillin.

The identification of DNA as the genetic material in 1944* and the rapid development of bacterial genetics starting about 1950 were the beginnings of a great new wave of discoveries that opened unexpected vistas in microbial biotechnology. It is of historical interest that the brilliant 1944 discovery that genes were composed of DNA was generally ignored for some time. One of the most outstanding contemporary microbial geneticists, William Hayes, has commented that in the mid-1950s, "I myself met a number of geneticists about this time who did not believe that the genetic material was DNA."

Nevertheless, it soon became clear that a large variety of bacterial mutants could be readily isolated from populations exposed to agents that affect the structure of DNA. Some of these were of special interest for biotechnology, namely, strains in which the mutation altered normal metabolic controls, leading to overproduction of a useful enzyme or organic chemical. In mutants of this sort, metabolism is "deranged" or, one could say, "unbalanced." It is often possible to devise culture con-

---

*The fact that genes are made of DNA was established in 1944 from experiments with pneumococci, bacteria that cause pneumonia. Maclyn McCarty, one of the three microbiologists involved, recently published a fascinating account of how this major research advance was made: McCarty, M., 1985, *The transforming principle; discovering that genes are made of DNA*. W. W. Norton, New York.

ditions that permit such mutants to grow and perform as if they were chemical factories designed to make products for human consumption. The food flavor-enhancing substance MSG (monosodium glutamate) is obtained in this way, using a bacterial metabolic "freak."

It is now clear that the development of recombinant DNA methodology in 1973 marked the beginning of another tidal wave of basic discoveries that have great promise for biotechnological applications. In anticipation of these, large chemical and pharmaceutical companies have made sizable investments of money and personnel for exploration of new products useful for treatment and control of infectious and other diseases, for agricultural productivity, and for other purposes. In addition, hundreds of new, relatively small, biotechnology companies have been formed in pursuit of the same goals.

There is general agreement that recombinant DNA technology is capable of producing many new and useful drugs, industrial solvents, fertilizers and so on. On the other hand, significant questions about the dangers of this new technology persist, such as

- Can we definitively exclude the possibility that new genetically engineered microbes may inadvertently become agents of "new" diseases?
- Could the recombinant DNA technology be used to devise dreadful agents of biological warfare?
- Could release into the environment of genetically engineered microbes (for example, designed to improve an agricultural practice) have unexpected and undesirable ecological effects?

These and related questions have raised complex issues affecting public policies and the social responsibilities of scientists. The history of debate on these matters and current views on the control and future development of recombinant DNA technology are summarized in *The Gene-Splicing Wars (Reflec-*

tions on the Recombinant DNA Controversy), R. A. Zilinskas and B. K. Zimmerman (editors), Macmillan, New York, 1986. The following remarks from a newsletter distributed by Congressman Lee Hamilton (9th District, Indiana; January 16, 1985) are relevant:

> Regulation of the emerging biotechnology industry is an important challenge facing the 99th Congress. Genetic engineering and other related forms of biotechnology are viewed by some as the most promising frontier since computers, offering us everything from double-sized livestock to a cure for cancer. Others, frightened by the possibility that man-made organisms may wreak havoc on the environment, see biotechnology as a major menace. Yet there is general agreement among ecologists and biotechnicians alike that federal regulation of the fledgling industry is necessary. How tightly controls should be drawn is the question. . . .The regulatory debate is taking place at the outset of biotechnology's development. Increased public attention to the issue means that we can hope to conduct a thorough examination of options and find a balanced solution to the problem.

In connection with Hamilton's comments it is of interest that in 1986, the Danish Parliament passed "The Law on Gene Technology and Environment." It is now unlawful to release genetically engineered organisms into the environment in Denmark, except in special instances approved by the Danish Minister for Environment. There is, of course, no way of preventing dust particles bearing microbes or bacteria on the feet of a bird from crossing borders. On the legal front in the United States, patent law specifies that to patent a "new" microbe or a process involving such a microbe, the organism must be deposited in a recognized culture collection. The U.S. Patent Office also requires assurances from depositors and repositories that the patent culture will be in the public domain permanently. At present, the American Type Culture Collection (ATCC) maintains over 2000 cultures of patented microbes. In addition to enormously increased paperwork, the staff of the ATCC will no doubt be faced with new varieties of technical and perhaps legal burdens.

22

Coda: Microbes and Early Life on Earth

The Earth itself is about 4 to 5 billion years old, but several lines of evidence indicate that life on Earth began approximately 3 to 3.5 billion years ago in the form of anaerobic bacteria. Geologists tell us that oxygen gas did not appear in the atmosphere until about 2 billion years ago, setting the stage for evolution of higher forms of life. Tracing the evolution of higher forms is aided greatly by the study of fossils, but fossils of early microbes are very rare and hard to find. New advances in biochemistry and molecular biology, however, have provided new tools for investigating the early evolution of microbes through detailed analysis of macromolecules in contemporary bacteria. It is believed that nucleic acid and protein macromolecules contain "molecular fossils," that is, atomic configurations that are relics of past evolutionary history. Using this information, molecular detectives should eventually be able to solve some of the mysteries of early evolution. We will then have a better understanding of how the many kinds of extant microbes are related to one another.

THE ORIGIN OF LIFE

The origin of life on Earth is one of the major unsolved mysteries of science. About 100 years ago, a famous Swedish chemist, Svante Arrhenius (1859–1927) popularized the idea that life did not originate on Earth, but came from elsewhere, presumably in the form of bacteria or bacterial spores. This notion, dubbed "Panspermia," was dismissed by thoughtful scientists some time ago because it does not explain anything; rather it merely shifts the problem to some other world. Until life of some kind is found elsewhere in our solar system, there is not much point in thinking about whether or not hypothetical microbes (or their spores) could have survived the interminable and hazardous journey to our planet. Curiously, some otherwise responsible scientists have recently begun to promote discussion of the possibility of Panspermia again, without adding new information of consequence.

The concept that a unique series of chemical and physical events resulted in the appearance of living procaryotes on the Earth more than 3 billion years ago is supported by evidence gathered from diverse fields of study. There is now reasonably general agreement on some important milestones:

Milestone	Billions of years ago
Age of the Earth	4.5
Age of the oldest terrestrial rocks	3.8
Age of the oldest known fossils of microbes found	3.5
First appearance of oxygen gas in the atmosphere	2.0
First appearance of eucaryotic cells	1.3–1.5

The following scenario is considered to be plausible by many scientists.

The first organisms on Earth were *anaerobic* procaryotes, and these were the only forms of life on the planet for a very long period. The Earth's atmosphere was devoid of oxygen gas until about 2 billion years ago. Consequently, early organisms could obtain energy only from anaerobic processes, fermentation of organic compounds in particular. (It is relevant to note at this point that fermentation is the simplest type of energy conversion process known.) With the passage of time, evolutionary improvements (due to mutations and other phenomena) gradually gave rise to cells with modified and more efficient bioenergetic systems. This eventually yielded anaerobic bacteria that could use light as their source of energy. Subsequent changes resulted in a crucial modification of the photosynthetic process, namely, the appearance of bacterial species that performed *oxygenic* photosynthesis. One evolutionary branch from such species led to green plants, which now produce virtually all of our atmospheric oxygen. As oxygen accumulated in the atmosphere, starting about 2 billion years ago, the stage was set for the evolution of *aerobic* bacteria and all the higher forms of life that depend on aerobic respiration for their energy needs.

The foregoing does not consider the fundamental question of how eucaryotic cell lines were first established, and this is still shrouded in mystery. There are several alternative possibilities, and new approaches are being pursued to unravel the evolutionary relationships of different species of procaryotes to one another and to eucaryotes. One popular current line of experimentation is the determination of base sequences in ribonucleic acids (RNA). Several kinds of RNA molecules occur in all living cells, and they participate in the actual mechanics of protein synthesis. The blueprints of amino acid sequences in proteins are encoded in DNA, whereas RNA molecules are involved in the mechanism of hooking amino acids together. DNA and RNA are molecular "cousins." They have some similar features but differ in that:

1. RNA is usually single-stranded,
2. the five-carbon sugar of RNA is slightly different from the corresponding sugar of DNA, and

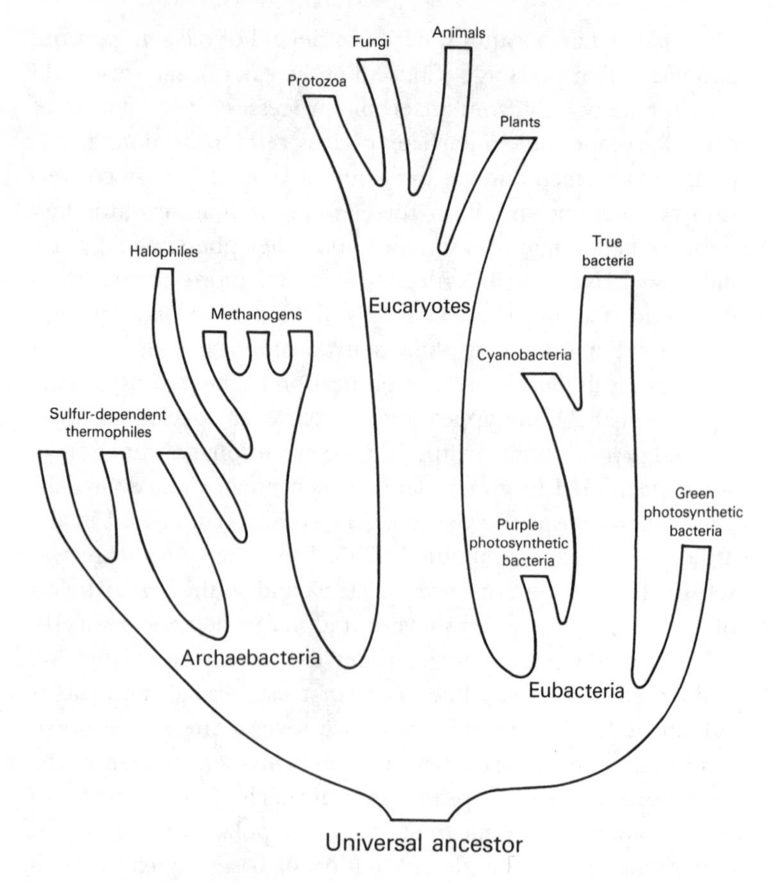

Figure 40 A phylogenetic (evolutionary) tree, modified from C. R. Woese. This interpretation of the evolutionary tree is based on comparisons of the chemical structures of an important class of RNA molecules found in all living cells.

3. one of the four nitrogen-containing bases of RNA has a different structure.

Some molecular biologists believe that the best molecular fossils of past evolutionary history will be found in the information present in base sequences of certain classes of RNA molecules. Intense study of such sequences in a variety of organisms over the past decade has led to the suggestion that procaryotes should be reclassified into two separate groups as shown in the evolutionary tree shown in Figure 40. It is suggested that all living organisms fall into one of three categories: eucaryotes, eubacteria, or archaebacteria. *Archaebacteria* are special bacteria that share certain biochemical features absent from other kinds of bacteria. For example, methanogenic bacteria and certain thermophiles are prominent representatives of archaebacteria. Bacterial species *not* included in this category are designated as *eubacteria.* Despite the ill-advised prefix, there is no evidence that archaebacteria are more ancient than eubacteria. Although RNA base sequences have provided new information of interest, it seems unlikely that a single biochemical "Rosetta Stone" can reveal the evolutionary relationships of the enormously diverse collection of microbes on Earth. We can expect, however, that our cornucopia of microbes will continue to offer many new opportunities for solving the basic riddles of life processes and for understanding how these processes changed during the early history of our planet.

Appendix I
How Leeuwenhoek Estimated the Sizes of Microbes*

As they'll say 'tis not credible that so great a many of these little animalcules can be comprehended in the compass of a sand-grain, as I have said, and that I can make no calculation of this matter, I have figured out their proportions thus, in order to exhibit them yet more clearly to the eye: Let me suppose, for example, that I see a sand-grain but as big as the spherical body ABGC [Text-fig. 3, p. 213] and that I see, besides, a little animal as big as D, swimming, or running on the sand-grain; and measuring it by my eye, I judge the axis of the little animal D to be the twelfth part of the axis of the supposed sand-grain, AG; consequently, according to the ordinary rules, the volume of the sphere ABGC is 1728 times greater than the volume of D. Now suppose I see, among the rest, a second sort of little animals, which I likewise measure by my eye (through a good glass, giving a sharp image); and I judge its axis to be the fifth part, though I shall here allow it to be but the fourth part (as Fig. E), of the axis of the first animalcule D; and so, consequently, the volume of Fig. D is 64 times greater

*From his letter of November 12, 1680, to the Royal Society of London: *in* Dobell, C., 1932, *Antony van Leeuwenhoek and His Little Animals*. Staples Press, London. Further details can be found in Schierbeek, A., 1959, *Measuring the Invisible World; the Life and Works of Antoni van Leeuwenhoek FRS*. Abelard-Schuman, London/New York.

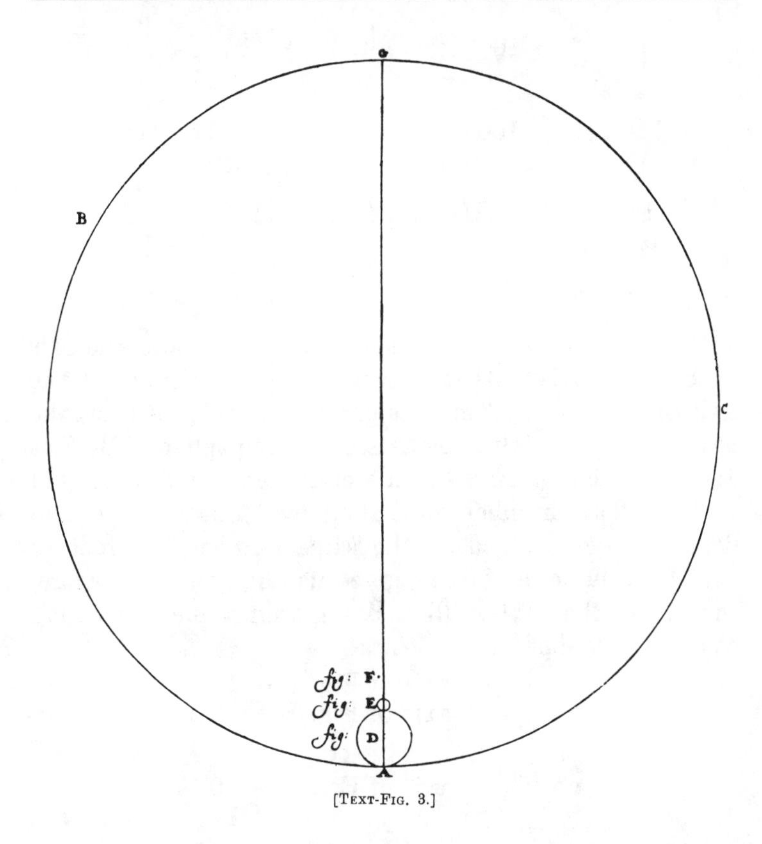

[Text-Fig. 3.]

than the volume of Fig. E. This last number, multiplied by the first number [1728], comes then to 110592, the number of the little animals like Fig. E which are as big (supposing their bodies to be round) as the sphere ABGC. But now I perceive a third sort of little animalcule, like the point F, whereof I judge the axis to be only a tenth part of that of the supposed animalcule E; wherefore 1000 animalcules such as F are as big as one animalcule like E. This number, multiplied by the one foregoing [110592], then makes more than 110 million little animals [like F] as big as a sand-grain.

12	10	4	1728
12	10	4	64
144	100	16	6912
12	10	4	10368
288	1000	64	110592
144			1000
1728			110592000

Otherwise I reckon in this fashion : Suppose the axis of Fig. F is 1, and that of Fig. E is 10; then, since the axis of Fig. D is 4 times as great as that of Fig. E, the axis of D is 40. But the axis of the big sphere ABGC is 12 times that of Fig. D; therefore the axis AG is equal to 480. This number multiplied by itself, and the product again multiplied by the same number, in order to get the volume of ABGC, gives us the result, as before, that more than 110 million living animalcules are as big as a grain of sand:

axis of Fig. F =	1
axis of Fig. E =	10
	4
axis of Fig. D =	40
	12
	80
	40
axis AG =	480
	480
	38400
	1920
	230400
	480
	18432000
	921600
	110592000

Appendix II
Some Microbes Added to the American Type Culture Collection During the Past Decade*

Organism	Outstanding characteristic
BACTERIA	
Acetobacter cellulolyticus (33288)	Degrades cellulose
Acetobacter pasteurianus (23754)	Acidifies vinegar by producing acetic acid
Aquaspirillum magnetotacticum (31632)	Accumulates magnetite and responds to a magnetic field
Arthrobacter petroleophagus (21494)	Used for production of single-cell protein from gaseous hydrocarbons
Bacillus acidocaldarius (27009)	Isolated from acidic thermal environments; can grow at temperatures from 45 to 70°C (113 to 128°F) and in the pH range from 2 to 6
Bacillus larvae (25748)	Produces disease in honeybee larvae, thereby impairing honey production
Bacillus sphaericus (33203)	Used for biological control of mosquitoes

*The numbers in parentheses are catalogue numbers used for specifying the organisms.

Organism	Outstanding characteristic
Bacillus thuringiensis (33679)	Produces disease in insect larvae; used for biological control of insect pests
Ectothiorhodospira vacuolata (43036)	Purple photosynthetic bacterium (isolated from a salty swamp at El Azraq, Jordan)
Erwinia ananas (31225)	May be useful for reduction of frost damage to plants
Heliobacterium chlorum (35205)	Green photosynthetic bacterium that contains a hitherto unknown form of chlorophyll (isolated from soil in front of Jordan Hall of Biology, Indiana University, Bloomington)
Lactobacillus sanfrancisco (27651)	Used for production of San Francisco sourdough bread
Legionella pneumophila (33152)	Caused Legionnaire's disease in Philadelphia
Leuconostoc mesenteroides (27258)	Produces dextran from sucrose; used for desugaring eggs
Mycoplasma genitalium (33530)	Causes urinary tract infections
Mycoplasma hypopneumonia (25095)	Causes pneumonia in swine
Spiroplasma citri (29051)	Causes corn stunt disease
Thiobacillus ferrooxidans (19859)	Used for removal of sulfur from coal
Thiobacillus thiooxidans (21835)	Used for cleaning metallic surfaces
FUNGI	
Beauveria bassiana (originally *Botrytis bassiana*) (48023)	Used for biological control of potato beetle
Culicinomyces clavosporis (46257)	Used for biological control of mosquitoes
Mucor hiemalis (20020)	Used for cleaning pearls (removes a natural yellow pigment)
Paecilomyces carneus (28276)	Used to remove phosphate in sewage treatment
Saccharomyces bailii, var. *osmophilus* (28166)	Causes spoilage of table wine

Appendix III
Microbes in Early Science Fiction

The subject of immunity against microbes was a key element in one of the great classics of science fiction, H. G. Wells' *The War of the Worlds*, published in 1898. This imaginary war was waged by inhabitants of Mars against our earthlings. Why did Wells choose Martians as our antagonists? Using ordinary telescopes, astronomers in the nineteenth century observed what appeared to be seasonal changes of color on Mars and of darkness of certain straight lines on the surface. To some, these lines looked like channels indicating the work of intelligent beings; this belief helped to make Mars an attractive setting for science fiction stories. In Wells' novel the Martians were shot to Earth in giant cylinders that landed in the English countryside. Wells' narrator describes the terrible destruction caused by the Martians; they operate gigantic mechanical monsters equipped with a "Heat-Ray" and other weapons. As the Martians approach London, panic begins, and a dispatch from the Commander-in-Chief of the British armed forces announces in the London newspapers:

> The Martians are able to discharge enormous clouds of a black and poisonous vapour by means of rockets. They have smothered our batteries, destroyed Richmond, Kingston, and Wimbledon, and are advancing slowly towards London, destroying everything on the way. It is impossible to stop them. There is no safety from the Black Smoke but in instant flight.

Eventually the narrator is able to observe the Martians from a concealed hideaway in the debris of a ruined house.

> They were, I now saw, the most unearthly creatures it is possible to conceive. They were huge round bodies— or, rather, heads—about four feet in diameter, each body having in front of it a face. This face had no nostrils—indeed, the Martians do not seem to have had any sense of smell, but it had a pair of very large dark-coloured eyes, and just beneath this a kind of fleshy beak. In the back of this head or body—I scarcely know how to speak of it—was the single tight tympanic surface, since known to be anatomically an ear, though it must have been almost useless in our dense air. In a group round the mouth were sixteen slender, almost whiplike tentacles, arranged in two bunches of eight each. . . .Strange as it may seem to a human being, all the complex apparatus of digestion, which makes up the bulk of our bodies, did not exist in the Martians. They were heads—merely heads. Entrails they had none. They did not eat, much less digest. Instead, they took the fresh, living blood of other creatures, and *injected* it into their own veins. I have myself seen this being done, as I shall mention in its place. But, squeamish as I may seem, I cannot bring myself to describe what I could not endure even to continue watching. Let is suffice to say, blood obtained from a still living animal, in most cases from a human being, was run directly by means of a little pipette into the recipient canal. . . .

London becomes a dead, silent city and then, suddenly, the Martians seem to begin disappearing. The narrator discovers what has happened after climbing an earthen rampart on the side of the Martian's headquarters:

> A mighty space it was, with gigantic machines here and there within it, huge mounds of material and strange shelter places. And scattered about it, some in their overturned war-machines, some in the now rigid handling-machines, and a dozen of them stark and silent and laid in a row, were the Martians—*dead*!—slain by the putrefactive and disease bacteria against which their systems were unprepared. . . .For so it had come about, as indeed I and many men might have foreseen had not terror and disaster blinded our minds. These germs of disease have taken toll of humanity since the beginning of things—taken toll of our prehuman ancestors since life began here. But by virtue of this natural selection of our kind we have developed resisting power; to no germs do we succumb without a struggle, and to many—

> those that cause putrefaction in dead matter, for instance—our living frames are altogether immune. But there are no bacteria in Mars, and directly these invaders arrived, directly they drank and fed, our microscopic allies began to work their overthrow. Already when I watched them they were irrevocably doomed, dying and rotting even as they went to and fro. It was inevitable. By the toll of a billion deaths man has bought his birthright of the earth, and it is his against all comers; it would still be his were the Martians ten times as mighty as they are.

EPILOGUE

In 1938, Orson Welles produced a radio program based on H. G. Wells' *The War of the Worlds* in the form of a simulated newscast, describing an invasion of New Jersey and New York by Martians. Many listeners apparently missed or did not listen to the introduction and subsequent announcements that clearly stated the fictional nature of the program. Welles' "newscast" caused near panic in many communities in New York, New Jersey, and elsewhere, as reported in the *New York Times* account of October 31, 1938.

The possibility of life on Mars persisted in the minds of some scientists into the 1960s. By 1969, new data on the composition of the Martian atmosphere and crust indicated that conditions on the planet would be very hostile to life as we know it, especially because water does not exist on the surface of Mars in liquid form. Moreover, the average temperature on the Martian surface is −55°C (−67°F)! In 1971, however, photographs made by Mariner fly-by spacecraft showed natural channels on the Martian surface that looked as if they had been cut in the past by running water. This observation led to NASA's Viking missions, whose major objective was to run automated tests on the surface of Mars for the presence of life in any form. Two U.S. spacecraft landed on Mars in 1976. A recent book* by

*Horowitz, N. H., 1986, *To Utopia and Back: The Search for Life in the Solar System*. W. H. Freeman, San Francisco.

Norman Horowitz, former head of the Jet Propulsion Laboratory's bioscience section for the Mariner and Viking missions, describes and evaluates the tests. There were three main results. First, cameras showed immediately that there were certainly no living organisms present of a size greater than several millimeters. Second, chemical tests by instruments on the spacecraft showed that the Martian surface contains no organic matter at a "parts-per-billion level of detectability." Third, three kinds of instruments designed to detect metabolic activities of microbes gave negative test results. Horowitz concludes:

> Viking found no life on Mars, and, just as important, it found why there can be no life. Mars lacks that extraordinary feature that dominates the environment of our own planet, oceans of liquid water in full view of the sun; indeed, it is devoid of any liquid water whatsoever. It is also suffused with short-wavelength ultraviolet radiation. Each of these circumstances alone would probably suffice to ensure its sterility, but in combination they have led to the development of a highly oxidizing surface environment that is incompatible with the existence of organic molecules on the planet. Mars is not only devoid of life, but of organic matter as well.

It is of interest that the possibility of "terrestrial-like" life on Mars was carefully considered by Philip Abelson in a prescient analysis* which indicated that it was quite unlikely. In his words: "If life actually exists on Mars it cannot be like any terrestrial form of life because of the relative absence of water. The crucial difficulty is the inability of life to function in a nonfluid state."

Wells was right, after all, about the absence of microbes on Mars. Is it possible that there are microbes on the moon? Horowitz notes: "The samples brought from the moon by the Apollo

*From Abelson, P. H., 1961, Extra-terrestrial life. *Proceedings National Academy of Sciences, U.S.*, 47:575.

crews have been studied more carefully from more viewpoints by more different scientists in a more organized way, perhaps, than any materials ever investigated. All tests for living organisms have been negative."

Appendix IV
The Ingenious Use of Microbiology Under Adverse Conditions*

MICROBIOLOGICAL EXPERIENCES IN JAPANESE CAMPS FOR PRISONERS OF WAR

by

G. GIESBERGER

(Received October 31, 1946).

> Bacteriology has not merely unveiled to the human mind unknown and unconjectured aspects of the microscopical world, it has moreover furnished means for the mastering of the benefits this microscopical world can offer and for the controlling of the misery which it can cause. It has tremendously extended our power over the phenomena of life.
>
> M. W. BEIJERINCK. Inaugural address, Delft. Sept. 26th 1895.

The enthusiastic words in which the great microbiologist BEIJERINCK tried to reveal the importance of his branch of science have, if ever, found their response in the conditions occurring in the numerous Japanese camps for prisoners of war and for internees during the occupation. Against the great misery caused by various infectious diseases, which through insufficient means were hard to combate, many cases exist, where the inhabitants of the camps

---

*From Giesberger, G., 1947, *Antonie van Leeuwenhoek Journal of Microbiology and Serology*, 12:267. The author, a Dutch microbiologist, was imprisoned in camps on the island of Java, Indonesia, during World War II.

were glad to make use of the special potencies of various microorganisms for the improvement of conditions they were living in.

My endeavours in the microbiological field in the camps already date from the first months. In order to surmount the difficulties met with in the baking of bread with the available sour dough, it was tried to isolate a yeast species, which might be cultivated on a large scale for the baking. The isolation actually succeeded by streaking the sour dough on a slice of a ripe papaja, as nutrient agar was lacking. A flat well-closed tin box served as petri dish. The yeast species present in the sour dough thrived so well on the slice of papaja, that after some transfers a pure culture was obtained. As the camp was abolished shortly after and the next one was provided with baker's yeast by the Japanese, a further development on a large scale has not been realized.

New and strong demands for applied microbiology in the new camp (4th and 9th Depot-Batallion at Tjimahi) came from medical quarters. Many inhabitants suffered from a more or less continuous diarrhoea, due to gastroenteritic disturbances caused by the abnormal nourishment. Several physicians applied chalk as an astringent, which, however, was only available in very small quantities. As lime stone was found by chance in large quantities in this camp and when burnt produced a lime of good quality, it was resolved to prepare precipitated chalk from this lime. For the conversion of $Ca(OH)_2$ into $CaCO_3$, carbon dioxide was used which was produced by a microbiological process.

The remains of rice from the kitchen, which in that period were still available in sufficient amount, served as raw material. Later, when hardly any rice was available,. the remains of bread were used successfully. The rice was subjected to a process of saccharification by means of the fungus *Rhizopus oryzae*, occurring in „ragi cakes''. The thus obtained sweetish tasting paste (known as „tapé'' to the native population) after having been mixed with water was left to a spontaneous fermentation in iron tanks of $\pm$ 100 l (derived from disused kitchen wagons). Yeasts as well as lactic acid bacteria took part in this fermentation process. The fermentation of a single lot took about two days. By means of this fermentation process during many months a continuous, quite satisfactory production of carbon dioxide was arrived at (not rarely over 1 l per minute). In this way chalk of good quality was obtained. It not merely served medical ends but as well was used as the raw material for a tooth paste prepared in the camp. During nearly one year, the period in which this branch of industry worked, some hundreds of kilograms of dried chalk and several thousand portions of tooth paste have been produced.

As the whole system was based on the microbiological conversion of rice amylum by *Rhizopus oryzae* (comparable with the so-called amylo process of the breweries) and as in many cases the preparation of food yeast was based upon it, I shall summarily mention some questions bearing on this process. Initially the rice has been inoculated with finely ground ragi-cakes, which were obtained from outside the camp. The native population prepare these cakes out of dough from rice flour in which the fungus develops spontaneously and forms spores. In dried condition the cakes contain the fungus so to say in conserved condition. Later spore suspensions have served as inoculum of the fungus. These were obtained by storing a moistened empty rice bag, which will always contain some rests of rice, under moist conditions. After a few days a thick fungus growth forms on the bag, first of a dirty white, but soon of a blueish black due to the numerous spores. These spores appeared ideal for the induction of a rapid and constant saccharification of the rice. The condition of the steamed rice appeared of major importance for the succeeding of the saccharification process. In fact when the rice is too moist and sticky, the process may fail completely. Instead of the favourable fungus, bacteria and yeasts develop, inducing a rapid rise in temperature and finally the rice turns into a disagreeable, musty smelling, slimy mass. Good aeration and not too high a temperature (of course no self-heating may occur) are essential. Especially when the remains of rice from the kitchen, which were often strongly contaminated, had been put to use, difficulties were often met with. When, however, freshly steamed rice from the kitchen had been used, the process took nearly always a favourable course. In 2—3 days the saccharification had proceeded sufficiently for the complete conversion of amylum into dextrin, maltose and glucose. A sweetish liquid dripped from the sticky mass, which after thickening by evaporation could furnish a native tit-bit, the „brem''.

Of much greater importance, however, than this microbiological source of chalk, was the production of a „food yeast''. Numerous inhabitants of the camp, as a result of the inappropriate nourishment suffered more or less seriously from Vitamin-B deficiency. As in that period baker's yeast could be had at a not too exorbitant high price from outside the camp in Tjimahi, cheap raw material was needed would it be economically allowed to culture one's own yeast. Again rice (later bread) remains of the kitchen were appropriate for this use. Two methods existed for the putting to use of this material after saccharification by *Rhizopus oryzae*. The first one consisted in preparing a wort from the saccharified rice by diluting with water, inoculating with yeast and, when the fermentation had come to a close, providing the patients with this wort as a whole. This system might be claimed as the most suitable under conditions such as they existed in most of the camps. In fact in Ambon, Batavia and Singapore etc. this method has been followed starting from fresh rice from the kitchen. It has been tried, prior to the inoculation with the yeast, to sterilize the wort as far as possible; by lack of fuel, however, this could not always be realized.

In Tjimahi, however, the remains of rice to be worked with were often contaminated in such a measure, that consumption of a yeast containing wort, prepared from such material was hardly tempting. Another method in this case led to a satisfactory solution. In fact it appeared that after fermentation of the wort from tapé, such as this took place in the CO_2 tanks for the production of chalk, on the surface of the fermented wort left in the air a definite yeast species developed spontaneously and abundantly. This yeast, (perhaps a *Torula* species, as in the literature on yeasts in tapé a species of *Torula* is mentioned as normally occurring) appeared to find favourable growth conditions in the oxidising of alcohol, lactic acid etc. occurring in this medium. In order to favour the development later some dedek (bran from the rice mills mixed with the pericarps) has been added to the medium. By simple means the yeast could be obtained in fairly pure condition by skimming the thick, rimpled pellicle off the surface followed by a further purification by means of repeated washings. In this way a thick yeast somewhat smelling of cheese was obtained which in consistency did not differ much from baker's yeast. The yield was very high, the more so after a special race with large cells and high production had been selected. From 100 l steamed rice about 15 kg of yeast could be obtained. When by means of control tests by a number of persons it had been certified that the yeast had about the same value as pressed yeast in combating the phenomena of Vitamin B deficiency, it was resolved to cultivate this yeast on as large a scale as possible. Thus fermentation basins with a large surface were constructed in which the tapé wort was poured in a thin layer of 10—15 cm. A shed with a cement floor served very well for the instalment of the fermentation basins, their walls consisting of wooden boards. These boards were first cemented on ridges in the floor. Later, when cement was lacking the boards were made leak proof by plastering them on the outer surface with clay from the rice fields. When this industry was at its full height the total surface of the yeast tanks came up to 75 m^2. The layer of the yeast in the basin was daily skimmed off, leaving merely a thin film and after 24 hours the yeast had again developed into a thick layer. From each basin yeast could be harvested during 7—10 days, depending on the degree of saccharification of the rice. Many hundreds of inhabitants of the camp were thus provided with a fair quantum of yeast, in pasteurised condition, supplied with sugar and cinnamon for the improvement of the taste. A part of the residual fairly acid wort from the yeast tanks was finally used in the manufacturing of paper for the breaking up of the fibres of the waste paper during the boiling.

Leaving the cultivation of yeast I will further point to the fact that in many camps alcohol has been produced for medical ends (secretly also for consumption) by means of fermentations of solutions containing sugar. The destillation of alcohol through lack of material was often realized by means of wondrously improvised destillation apparatus. In Tjimahi for instance I made use of a large enamelled teakettle as a still, the spout being connected with the cooler which next to a water jacket out of tin contained as essential part the glass stringtube of a well-known violinist.

In Tjimahi the thus obtained alcohol was moreover used for the preparation of vinegar for use in the kitchens as soon as this payed, the vinegar which could be obtained from outside the camp being high in prize and very low in quality. The alcohol, however, was not conducted over chips of wood but over long strips of bag cloth strained parallel one to the other in a wooden chimney. When after some lapse of time a flora of acetic acid bacteria had developed (among which *Acetobacter xylinum* occurred regularly), the conversion of alcohol into acetic acid took a rapid course. The correct moment for bringing the process to a stop was determined titrimetrically. In this connection it may be observed that in the absence of any indicators, natural indicators had to be looked for. A red substance in the marrow of the root of the so-called kajoe setjang appeared to serve this end successfully, while later a dark red substance occurring in the root of a species of mangrove was appropriate as well.

As a final example of the application of microbiology the preparation of ,,tempé" from soybeans may be cited. Experience learns that soybeans as such are hard to digest. The native population have always prepared a much valued, tasty and nourishing product from soybeans, the ,,tempé kedelee" by letting boiled or steamed beans grow mouldy. In many cases it is *Rhizopus oryzae* or a nearly related fungus which plays its parts here and which by its development induces a sufficient breaking up and thus a better digestibility. It stands to reason that the preparation of tempé was taken up in the camps in as far as soybeans were available. Difficulties have been often met with, however, because the desired fungus did not or insufficiently develop and putrefying bacteria took the leading instead. By slightly acidifying the soybeans, the development of the bacteria was inhibited, but the actual factors benificent for moulding consisted in a good aeration and the maintaining of a not too high temperature. Care had to be taken that the bean mass was not subjected to the moulding process in too moist a condition. After this experimental evidence in most of the camps the production of a tempé of sufficient quality was arrived at and thus the inhabitants of the camp could enjoy the benifit of this much valued food stuff.

I have now reached the end of this survey of the principal applications of microbiology in the Japanese camps for prisoners of war. If most of the camps had not been removed over and over again, doubtless much more might have been accomplished in this field. So in Tjimahi plans existed in an advanced state for the preparation of ammonia from urine by means of urea bacteria in order to produce a lye for the manufacturing of soap. When, however, the total sum is considered there is in my opinion every reason for satisfaction with the results obtained, which without the widened field of view which applied microbiology offers us, would never have been realized.

Who's Who in This Book

Bassi, Agostino. *1773–1856.* Italian lawyer and agriculturist. After holding various positions in the civil service, Bassi devoted his efforts to analysis of agricultural problems and related matters. He established that the causative agent of the silkworm disease muscardine was a miscroscopic fungus. This was the first demonstration that an infectious animal disease was due to a microbe (1835). In his later writings, Bassi suggested that "parasites" were the causes of various diseases (including plague, smallpox, syphilis, and cholera) and advocated methods of prevention.

Escherich, Theodor. *1857–1911.* Well-known pediatrician and bacteriologist. Born in Munich. Practiced pediatrics in Munich. Professor in Graz, Austria and finally in Vienna. Discovered the organism *Escherichia coli.*

Fleming, Alexander. *1881–1955.* British bacteriologist. Born Lochfield, Scotland. Educated at St. Mary's Hospital Medical School at the University of London and returned to teaching there after serving in the army medical corps during World War I. Professor at the Royal College of Surgeons. Discovered penicillin in 1928. Admitted to the Royal Society in 1943, knighted in 1944, and awarded the Nobel Prize in 1945.

Fracastoro, Girolamo. *ca. 1478–1553.* Physician, astronomer, geographer, poet, and humanist. Born at Verona and studied in Padua. He was chief physician to the Council of Trent. He published his poem *Syphilis sive Morbus Gallicus* (1530) and *De sympathia et antipathia rerum* and *De contagione* in 1546.

Hooke, Robert. *1635–1703.* English experimental philosopher and mechanical genius. Born at Freshwater, Isle of Wight. Educated at Westminster and Oxford. Secretary of the Royal Society 1677–82. Hooke published his famous *Micrographia* in 1665 and did much to rouse interest in microscopy.

Jenner, Edward. *1749–1823.* English physician. Born at Berkeley. Obtained the M.D. from St. Andrews. Discovered the principle of vaccination as a result of his study of patients with smallpox and cowpox. In 1803, the Royal Jennerian society for the spread of vaccination in London was established. Conferred an honorary M.D. by Oxford in 1813.

Koch, Robert. *1843–1910.* Born in Clausthal, Hannover, Germany, the son of a mining engineer. Studied in University of Göttingen and graduated as doctor of medicine in 1866. Served as surgeon in Franco-Prussian war, and in 1872 became district medical officer in Wollstein. Published his classical research on anthrax in 1876, on the technical methods of bacterial examination in 1877, and on the etiology of traumatic infective diseases 1878. Became associated with the Imperial Health Office in Berlin and founded famous school of bacteriology there. In 1881 he solved the problem of pure bacterial cultures. Later his methods were universally employed. Discovered tubercle bacillus in 1882, and cholera vibrio in 1883. Travelled extensively studying protozoal diseases in Africa and India. Received Nobel Prize 1905. Foreign member of Royal Society 1897.

Leeuwenhoek, Antoni van. *1632–1723.* Great Dutch microscopist and first discoverer of bacteria. Born in Delft. In youth served as a bookkeeper in a draper's shop in Amsterdam, but returned to his native town when 22 years of age and died there aged 91. Remained in obscurity for 40 years, but discovered how to grind microscopic lenses and in 1673 was introduced to the Royal Society of London and became one of its most famous correspondents. He was elected a fellow of the Royal Society in 1680 and wrote about 200 letters to the Society, containing accounts of hundreds of discoveries which he had made using his lenses. He first saw living protozoa in 1674 and bacteria in 1675.

Metchnikoff, Élie. *1845–1916.* Russian zoologist, embryologist, and pathologist. Discovered the role of phagocytes in immunity. Educated at University of Charkow. Later studied at Giessen and Naples. Became professor and director of Bacteriological Institute in Odessa in 1886, but left in 1887 and went to Paris, where he resided till the end of his life. Assistant director of the Pasteur Institute. He received the Nobel Prize 1908.

Pasteur, Louis. *1822–1895.* Great French chemist and bacteriologist at the École Normale, Paris. Was professor of physics at the Lycée of Dijon 1848 and of chemistry in Strassburg 1852. Dean of the faculty of science at Lille 1854. Director of Studies in the École Normale, Paris. Pasteur carried out epoch-making researches demonstrating the connection between various fermentations and the activity of living microorganisms. Lactic fermentation, 1857; alcoholic fermentation, 1858–1860; butyric fermentation, 1861; acetic fermentation, 1861–1864. In 1877 he began the study of the causes and prevention of infective diseases in man and animals. In his honor the Pasteur Institute in Paris was founded 1888.

Priestley, Joseph. *1733–1804.* English chemist and minister who also published extensively on philosophy, education, religion, political theory, and the history of science. Born in Birstall. Earned a doctor of laws degree from the University of Edinburgh. Admitted as a fellow of the Royal Society in 1766. In 1767 completed the pioneering work *The History and Present State of Electricity*. Discovered oxygen and several other gases, including nitrogen, ammonia, and hydrogen chloride.

Tyndall, John. *1820–1893.* British physicist. Born in Ireland. Studied at Marburg and Berlin. Became professor at the Royal Institution in London 1853. Colleague of Faraday, whom he succeeded as Superintendent 1867-87. Wrote extensively on natural philosophy and was a popular lecturer and experimenter. In 1870 he began to interest himself in atmospheric germs and dust, and he carried out numerous exact experiments on sterilization by heat which led him to the discovery (1877) of fractional sterilization, now called Tyndallization. By his lectures and writings, Tyndall did much to further the teaching of Pasteur. He was accidentally poisoned and died 1893.

Volta, Alessandro. *1745–1827*. Italian physicist. Born in Como. Received education in classical studies through relatives who were members of the clergy. Published *De vi attractiva ignis electricii* 1769. Appointed teacher of physics and superintendent of the Royal School of Como (1774) and chair of physics at the University of Pavia (1779). In 1815 was made the head of the philosophical faculty at the University of Padua. Volta developed the notion of the electrochemical series, and his invention of the electric battery provided the first source of continuous current.

Suggestions for Further Reading

GENERAL MICROBIOLOGY

Postgate, J., 1986, *Microbes and Man* (2nd ed.). Penguin Books, Harmondsworth, England, 239 pages.

This interesting book is addressed to the general reader and surveys microbiology in nontechnical language.

Sistrom, W.R., 1969, *Microbial Life* (2nd ed.). Holt, Rinehart, and Winston, New York, 148 pages.

Sistrom emphasizes the biochemical activities of bacteria. The book is useful for students embarking on professional training for careers in microbiology or related sciences.

Wilkinson, J.F., 1975, *Introduction to Microbiology* (2nd ed.). Blackwell Scientific, Oxford, 120 pages.

A brief, simplified introduction to microbiology that assumes "the reader has an elementary understanding of the basic principles of biology and, in particular, those of biochemistry and, to a lesser extent, of genetics."

MICROBIAL ECOLOGY

Grant, W.D. and Long, P.E., 1981, *Environmental Microbiology*. Blackie, Glasgow, 215 pages.

This text was written for college undergraduates. The authors assume that readers will know something about the major microbial groups and are familiar with special features of bacteria, but intend the text for "biologists with only a rudimentary knowledge of microbiology." The book is divided into three sections entitled: habitats, microorganisms as environmental determinants, and microorganisms and pollution.

BIOTECHNOLOGY

Smith, J.E., 1981, *Biotechnology*. Edward Arnold (Publishers), London, (The Institute of Biology's Studies in Biology no. 136) 75 pages.

Smith describes biotechnology as "the science which studies the integrated application of microbiology, biochemistry, and process technologies on biological systems for their use in the interdisciplinary nature of science" and predicts: "Biotechnology will create wholly novel industries, requiring little fossil energy, and will be responsible for changing the world economy, particularly in the next century." This small book presents a clear, succinct, and informative account of current aims and procedures in modern biotechnology.

HISTORY OF MICROBIOLOGY

Brock, T.D., 1961, *Milestones in Microbiology*. Prentice- Hall, Englewood Cliffs, N.J., 275 pages.

This book presents a number of historically important research papers, many of them translated and edited by Brock. The time span is 1677 (van Leeuwenhoek) to 1940, and Brock provides helpful interpretive comments at the end of each paper.

McNeill, W.H., 1979, *Plagues and People*. Penguin Books, Harmondsworth, England, 330 pages.

A popular and eminently readable book by Hans Zinsser entitled *Rats, Lice, and History*, published in 1935, showed how epidemics of the infectious disease typhus influenced history. McNeill's scholarly book takes a new look at the dramatic impacts of various plagues and pestilences on political and social events of the past.

Reid, R., 1975, *Microbes and Men*. Saturday Review Press (A Division of E.P. Dutton), 170 pages.

This interesting and well- illustrated book was written to accompany a BBC television series on famous pioneers of infectious disease research.

GENERAL BIOLOGY

Starr, C. and Taggart, R., 1984, *Biology: The Unity and Diversity of Life* (3rd ed.). Wadsworth, Belmont, CA, ca. 700 pages.

Nowadays, biology texts for college freshmen are encyclopedic, and this is no exception. This text is very well crafted and beautifully illustrated. I have recommended it to nonscientist friends as a resource for the home library. The authors note that "Biology encompasses a tremendous and scattered array of subdisciplines, each with its squadrons of practitioners who are bent on discovery." Approximately 25 pages are devoted to the microbial world.

GENERAL CHEMISTRY

Rossotti, H. 1975, *Introducing Chemistry*. Penguin Books, Harmondsworth, England, 344 pages.

"Many of us have regrets that we were unable to study some subject at school, or that we abandoned it too readily." Dr.Rossotti wrote this book "for those who have such regrets about chemistry." A well-written "layman's account" of the fundamentals of chemistry.

GENERAL INTEREST

Roueche, B. 1980, *The Medical Detectives.* Times Books, New York, 372 pages.

Roueche is well known to readers of the New Yorker magazine for his fascinating articles on the "Annals of Medicine." This book is a collection of these articles published between 1947 and 1980. Each chapter is an actual case history written as a detective story. Many of the culprits are microbes or viruses. Exciting!

Medawar, P.B., 1981, *Advice to a Young Scientist.* Harper Colophon Books/Harper and Row, New York, 109 pages.

How do scientists go about making discoveries, propounding "laws," or otherwise enlarging human understanding? How can I tell if I am cut out to be a scientific research worker? Sir Peter Medawar, Nobel Laureate, and one of the most erudite of contemporary scientists, answers these and related questions and offers useful advice to both young and older scientists. Also to nonscientists "who may for any reason be curious about the delights and vexations of being a scientist, or about the motives, moods, and mores of members of the profession."

Glossary

Adenosine triphosphate See ATP.

Aerobe An organism that requires gaseous oxygen (O_2) for energy-yielding respiration.

Agar A polysaccharide obtained from seaweed; used as an agent for solidifying culture media.

Algae Photosynthetic eucaryotic organisms.

Amino acid Small nitrogen-containing molecule; amino acids are the building blocks of proteins.

Ammonification Conversion of the nitrogen in organic compounds to ammonia (NH_3).

Anaerobe An organism that does not require gaseous oxygen for energy metabolism; O_2 usually inhibits growth of anaerobes.

Antibiotic Chemical produced by a microbe, that effectively inhibits or kills other species of microbes.

Antibody Protein produced by white blood cells that reacts specifically with an antigen.

Antigen A substance such as a bacterial toxin, that induces formation of a specific antibody in animal blood or tissues.

Antimetabolite Chemical which inhibits or kills microbes.

Atom The smallest unit structure of chemical elements.

ATP An energy-rich molecule that contains three phosphate groups; referred to as the "energy currency" of cells.

Autotroph An organism that can use carbon dioxide as the sole source of carbon for growth.

Bacteriophage ("phage") A virus (parasite) that infects and multiplies in bacteria.

Base See *nucleic acid base.*

Biosynthesis The totality of the processes by which cells produce large molecules, such as proteins and cellular structures, from small chemical building blocks.

Biotechnology The application of scientific and technical advances in biology for manufacture of products useful in agriculture, medical treatment, or industrial processes.

Calorie The amount of energy required to raise the temperature of 1 gram of water from 14.5 to 15.5°C.

Carbohydrate An organic molecule, such as a sugar or polysaccharide, that contains carbon, hydrogen, and oxygen according to the general formula $C_n(H_2O)_n$.

Catalyst An agent that accelerates the rate of a chemical reaction, but remains unchanged (that is, it is not consumed).

Cell The basic organizational unit of living organisms.

Cellulose A polysaccharide composed of glucose units.

Chemical bond A linkage (force) by which two atoms are attracted or attached to each other.

Chlorophyll The green pigment found in photosynthetic organisms; it absorbs light, which is converted to chemical energy.

Clone A population of cells all of which were derived from a single ancestral cell.

Colony A population of cells arising from a single cell, growing on a semi-solid medium such as agar.

Compound Chemical substance composed of two or more different kinds of atoms held together by strong chemical bonds.

Conjugation A sexual reproduction process in which there is a transfer of genetic material between cells.

Contagious disease A disease which can be transmitted from an ill person to a healthy one.

Culture A population of cells grown under defined conditions.

Denitrification Anaerobic conversion of nitrate to nitrogen gas (N_2) catalyzed by certain microbes.

DNA Deoxyribonucleic acid, a type of nucleic acid that carries the genetic information that determines the characteristics of an organism; the 5-carbon-atom sugar component is deoxyribose.

Enzyme A protein that accelerates biochemical reactions in a catalytic manner.

Epidemic A disease found in a greater-than-usual number of individuals in a community at the same time.

Eucaryote A cell or organism that has a membrane-bound nucleus (includes all cell types except bacteria).

Fat A class of greasy organic substances found in all types of cells; fats are composed of glycerol combined with organic "fatty acids."

Fermentation Energy-yielding breakdown of organic compounds (such as sugars) in the absence of air (or oxygen).

Fungus A class of nonphotosynthetic organisms, some of which are microscopic (yeasts and molds).

Gene A segment of DNA that carries the genetic information needed for construction of a single protein.

Glucose A sugar with the formula $C_6H_{12}O_6$; also called dextrose or grape sugar.

Glycogen A polysaccharide made up of glucose units; a major storage product in animal cells (for example, muscle and liver) and some microbes.

Halophile A microbe that can grow in concentrated salt solutions

Heterotroph An organism that requires organic compounds as sources of energy and cellular carbon.

Host An organism capable of supporting the growth of a microbial parasite or virus.

Ion An atom or molecule that carries an electrical charge (for example, H^+).

Immunity Ability of an animal to resist infection.

Immunization Inoculation of humans or animals with bacteria, viruses, or other antigens for the purpose of provoking the production of protective antibodies.

Infection Growth of a microbe in a host which usually produces disease.

Inorganic compound "Nonliving"; A chemical substance which does not contain carbon (except for carbon monoxide and carbon dioxide).

Koch's postulates Experimental criteria proposed by Robert Koch to demonstrate that a specific disease is caused by a specific organism.

Macromolecule A large molecule, usually composed of numerous smaller chemical units.

Mesophile An organism that grows best at intermediate temperatures (20 to 45°C).

Metabolism The biochemical processes by which cells obtain energy and produce their characteristic constituents.

Micrometer Unit of length equal to one-millionth of a meter; μm.

Milliliter A measure of liquid volume equal to one-thousandth of a liter; essentially equivalent to the volume of 1 cubic centimeter; ml.

Molecule A combination of two or more atoms held together by chemical bonds.

Mutant strain A pure culture derived from a cell in which one or more mutations have occurred.

Mutation A chemical change in DNA leading to a change in a heritable genetic characteristic; some mutations are lethal.

Nitrification Conversion of ammonia to nitrate by microbes.

Nitrogen fixation Conversion of atmospheric N_2 gas to ammonia and cellular nitrogen compounds by microbes.

Nucleic acid A large molecule composed of subunits, each consisting of a nitrogen-containing part called the base, a 5-carbon-atom sugar, and a phosphate group.

Nucleic acid base A nitrogen-containing component of a nucleic acid.

Nucleus The membrane-bounded structure in a eucaryotic cell that contains genetic material in the form of chromosomes.

Organic compound A carbon compound that contains chemical bonds between carbon and hydrogen atoms (usually also other kinds of chemical bonds).

Parasite An organism or virus that lives on or in another organism (the host) and causes damage to the host.

Pasteurization A mild heat treatment of foods and beverages designed to kill pathogenic agents or microbes that cause spoilage.

Pathogenic Capable of causing disease.

pH A numerical measure of the relative acidity of solutions. The acidity value depends on the concentration of hydrogen ions (H^+), and the scale ranges from pH = 0 (most acidic) to pH = 14 (least acidic).

Phagocyte Type of white blood cell that can engulf and destroy microbes and other foreign bodies.

Photosynthesis The process in which light energy is converted to chemical energy and the latter used for the enzyme-catalyzed conversion of carbon dioxide to organic substances.

Plasmid A small ring of DNA that can replicate itself independently of the bacterial chromosome.

Polysaccharide A large carbohydrate molecule, such as starch and glycogen, consisting of many sugar units.

Procaryote A microbe (bacterium) that does not have a well-defined nucleus.

Protein A macromolecule consisting of amino acid units.

Protozoa Single-celled, nonphotosynthetic eucaryotes that have properties typical of animal cells.

Psychrophile A microbe that grows best at temperatures below 20°C.

Pure culture Population of microbial cells derived from a single cell.

Recombination The process in which genetic elements from two parent cells are brought together.

Respiration The oxygen-dependent process used by aerobes for obtaining energy.

RNA Ribonucleic acid, a type of nucleic acid that is involved in the cellular manufacture of proteins; the 5-carbon-atom sugar component is ribose.

Rumen The first stomach of a ruminant animal such as a cow, which is, in effect, an incubator in which protozoa and other microbes digest and degrade cellulose to small organic molecules by fermentation and related processes.

Spore A thick-walled cell produced by certain microbes, which is very resistant to heat, drying, and disinfectants.

Sterilization Any process that kills all microbes and viruses on and in objects.

Symbiosis An association of two organisms that involves some degree of interdependence and which is often mutually beneficial.

Thermophile An organism that grows best at temperatures above 45 to 50°C.

Toxin A poisonous substance produced by a microbe.

Vaccination Use of "vaccines" (microbes or viruses treated so as to remove their ability to cause disease) for the purpose of provoking protective antibody formation.

Virulence The degree to which a microbe or virus is disease-producing.

Virus An infectious agent of very small size that can multiply only inside animal, plant, or bacterial host cells.

Vitamin A special chemical required in the diet in very small amounts; specific vitamins function in association with particular enzymes.

Credits and Acknowledgments

Figure 1 Replica from the Archives of the American Society for Microbiology.

Figure 2 Leeuwenhoek, A. van., 1684, *Phil. Trans. Roy. Soc. London, 14*, 568.

Figure 4 C. H. W. Hirs *et al.*, 1960, *J. Biol. Chem. 235*, 633.

Figure 5 Brock, T., et al., 1986, *Basic Microbiology*. Prentice-Hall, New Jersey.

Figures 7a and b Courtesy of Dr. Thomas D. Brock, University of Wisconsin.

Figure 7c Courtesy of Dr. Norbert Pfenning, University of Konstanz, Germany.

Figure 11 Priestley, J., 1774, *Experiments and Observations on Different Kinds of Air*. Printed for J. Johnson, London. Photograph courtesy of the Lilly Library, Bloomington, Indiana.

Figure 14 Brock, T., et al., 1986, *Basic Microbiology*. Prentice-Hall, New Jersey.

Figure 15 Cartoon by Dr. Joachim Czichos, Heidelberg, Germany.

Figure 16 From Fuller, T., 1650, *A Pisgah-sight of Palestine and the Confines Thereof, with the History of the Old and New Tes-*

tament acted thereon. Printed by J.F. for John Williams, London. Courtesy of the Lilly Library, Bloomington, Indiana.

Figure 17 Photograph by D. Balkwill and D. Maratea; courtesy of Dr. R. Blakemore, University of New Hampshire.

Figure 18 Croome, R., and Tyler, P., 1984. *J. Genl. Micro. 130*, 2717.

Figure 19 Courtesy of Dr. David White, Indiana University.

Figure 20 Courtesy of Dr. R.A. Samson, Centraalbureau voor Schimmelcultures, Baarn, The Netherlands.

Figure 26 Lent by Mr. and Mrs. Arthur V. Brown II and Indiana Univ. Art Museum. Photograph by Ken Strothman and Harvey Osterhoudt.

Figure 32 Küster, E., 1915, *Arbeiten aus dem Kaiserlichen Gesundheitsamte, 48*, 1.

Figure 35 Fleming, A., 1929, *Brit. J. Exp. Path., 10*, 226.

Figure 39 Gonick, L. and Wheelis, M., 1983. *The Cartoon Guide to Genetics*, Barnes and Noble, New York.

Figure 40 Modified from Woese, C. R., 1984. *The Origin of Life.* Carolina Biological, Burlington, North Carolina.

Index

Acetobacter cellulolyticus, 215
Acetobacter pasteurianus, 215
Acetobacter xylinum, 226
Acetone, 203
Acid rain, 91
Acid, defined, 102
Acidophiles, 102–103
Acquired immune deficiency syndrome, 166
Activated sludge process, 141
Adenosine triphosphate, *See* ATP Aerobic microbes, 32
Aerobic respiration, 47–48, 67–69, 107, 209
Agar, 40–42
AIDS, *See* Acquired immune deficiency syndrome Airship Hindenburg, 74
Alchemists, 15
Alcohol, 15–17, 27, 97, 115, 118–120, 122, 203
Algae, 40, 124
Alkaline, defined, 103
Alvinella pompejana, 107
Alvinocaris lusca, 107
Alvin, U.S. Navy submersible, 106–107
American Type Culture Collection, 50–52, 206
Amino acids, 28, 49, 54, 86, 193–194, 209
Ammonia, 82–83, 85–87, 138, 141
Ammonification, 82–83
Anabaena, 125
Anaerobes, 14, 18, 139, 209
Anaerobic digesters, 76
Anaerobic respiration, 85, 92
Angkor Wat, 94
Antarctic lakes, 98
Anthrax, 52, 58, 60–61, 157, 165
Anthrax island, 60
Antibiotics, 52, 162, 183–186, 204
Antibodies, 162, 164–167
Antigens, 165–167
Antigen-antibody combination, 166–167
Antimetabolites, 182–183
Antiseptics, 182–183
Aquaspirillum magnetotacticum, 105, 215
Archaebacteria, 211
Archaeomicrobiology, 59–60
Aristotle, 100
Arrhenius, S., 208
Arthrobacter petroleophagus,215
Asexual reproduction, 37
ATCC, *See*American Type Culture Collection Atmosphere, microbes in, 35
ATP, 48, 69, 74, 86, 92–93, 116–120, 123, 126, 179
Auden, W.H., 168–170
Autoclave, 42
Autotrophs, 64–65, 74, 82, 107
Azotobacter vinelandii, 52

Bacillus, 43, 53, 59
Bacillus acidocaldarius, 215`
Bacillus anthracis, 52, 58, 60, 157, 177
Bacillus larvae, 215
Bacillus polymyxa, 52
Bacillus sphaericus, 215
Bacillus thermoproteolyticus, 97
Bacillus thuringiensis, 215
Bacteria-plant symbiosis, 89–90
Bacterial consortia, 109
Bacteriophages, 176–179, 188, 197
Baker, H., 57
Bardell, D., 6
Bartholomew, J.W., 59
Bassi, A., 13, 149–152
Beauveria bassiana, 216
Beer, 18
Behavior, magnetotactic, 104–106
Behavior, social, 108, 110–111
Beijerinck, M.W., 222
Bell, W.G., 147
Bergey's Manual, 11, 49
Biochemical mutants, 196–197, 198, 204
Biochemistry, 19
Biogas, 76–77, 141
Biological warfare, 60
Biotechnology, 47, 80, 88, 97, 165, 187–188, 199–206
Black Death, 146
Boghurst, W., 147
Bonds, chemical, 22–23
Bosch, C., 86
Botrytis bassiana, 152, 216
Botrytis paradoxa, 152
Botulism, 56
Brierley, C.L., 60
Brierley, J.A., 60
British Museum, 58
Bulloch,W., 147, 154
Burke, J., 57

Cagniard-Latour, C., 16
Caldwell, D.E., 60
Calorie, defined, 78
Carbohydrates, 25, 27, 31
Carbon cycle, 62–70
Carbon monoxide, 24, 70
Carlsberg Brewery, 18
Carotenoids, 124
Cartwright, F.F., 145, 158
Catalyst, definition of, 29
Cell preservation techniques, 51
Cells, sizes of, 3, 177
Cellulose, 26, 74, 76, 97, 114–115, 123
Chain, E., 185
Chamonix Glacier, 35
Chemical compounds, 21
Chemotherapy, 182
Chlorochromatium aggregatum, 110
Chlorophyll, 93, 123–125
Cholera, 37, 136, 152, 157, 162
Chromatium, 125
Chromatium vinosum, 65
Chromosome, 10, 188, 191–192, 197–198
Clams, giant, 107–108
Claviceps purpurea, 115
Cloaca Maxima, 136
Clone, 39, 41
Clostridium, 53
Clostridium acetobutylicum, 52
Clostridium botulinum, 52, 54, 56
Clostridium perfringens, 55
Clostridium tetani, 52, 58
Clostridium thermocellum, 97
Coal, 57, 70–71
Coenzymes, 130–132
Cohen, I.R., 195
Colony, 38–39, 41
Concrete, deterioration of, 94
Conjugation, 197–198
Convulsive ergotism, 115
Copper mining, 94
Cork, cellular structure of, 2
Cowpox, 159–160
Crick, F., 192, 196
Crookes, W., 85
Cross, T., 59

Cryptobiosis, 56–57
Culicomyces clavosporis, 216
Cyanide, 70, 94
Cyanobacteria, 65, 88, 124–125, 127

Dalton, J., 22
Dalton, weight unit, 22
David, life in germfree isolator, 173–174
Dead Sea, 100–101
Deep sea vents, 106–108
Denitrification, 83–85
Dental plaque, 109
Desulfobacter, 92
Desulfobulbus, 92
Desulfovibrio vulgaris, 92
Diarrhea, 46
Disinfectants, 182
DNA, 10, 25, 178–179, 187–202, 204–205, 209–210
Doctor's Dilemma, 183
Duclaux, E., 155
Dumas, J.B., 155
Dysentery, 162

Ecology, 75, 88, 95, 98, 105, 107–109, 125–126, 166, 169–170
Ectothiorhodospira, 104
Ectothiorhodospira vacuolata,216
Eddy, H.P., 137
Ehrlich, H.L., 11
Electron microscope, 176
Element cycles, 62
Elemental sulfur, 91, 93
Elements, chemical, 20–21
Endospores, 34–35, 53–61, 208
Endosymbiotic bacteria, 108
Energy, "currency" of (ATP), 48
Energy, from aerobic respiration, 48
Energy, from fermentation, 15
Energy, requirements, 31
Energy, reservoirs, 27
Enrichment cultures, 36–37
Enzymes, 26, 29–30, 130–131
Epsom salt, 92
Ergot, 115
Erwinia ananas, 216
Erwinia carotovora, 52
Escherichia coli, 46, 196–197, 199–202
Escherich, T., 46
Eucaryotes, 10, 113, 188, 209, 211
Eutrophication, 142
Evolution, 127, 207, 209–211
Exobiology, 220–221

Fats, 24–25, 30–31
Fermentation, 13–18, 27, 48, 55, 75, 97, 118–120, 122–123, 203, 209
Fleming, A., 183–185
Florey, H., 185
Food and Drug Administration, 56
Ford, B.J., 8
Fossil microbes, 207–208
Fossils, molecular, 207
Fracastoro, G., 148
Frankland, Mrs. P., 155
Frankland, P., 155
French Revolution, 67
Fuller, T., 100
Fungi, 10, 113–115, 184–185

Gas gangrene, 56
Gelatin medium, 38, 40
Gene cloning, 200–201
Gene transfer, 87–88, 188, 196–199
Genes, 188, 191–192, 194–196
Genetic code, 194–196
Genetic engineering, 47, 88, 188, 199, 205–206
Geomicrobiology, 11
Germ theory of infectious disease, 148
Germfree animals, 170–173
Giant clams, 107–108
Gieseberger, G. 222

Glucose, 22–27, 48, 70, 114–115, 118–120
Glycogen, 26–27
Great Salt Lake, 100
Green algae, 10
Gruinard Island, 60–61
Gulliver's Travels, 121
Gunpowder, 84
Gypsum, 92

Haber-Bosch process, 87
Haber, F., 86–87
Hales, S., 121
Halobacterium volcanii, 102
Halophiles, 99, 102
Hamilton, L., 206
Harden, A., 15
Hayes, W., 204
Heat of combustion, 78–79
Hedges, F., 155
Heliobacterium, 125
Heliobacterium chlorum, 52, 216
Helmont, J.B. van, 153
Hesse, F., 40
Hesse, W., 40
Heterotrophs, 65, 68
Hooke, R., 2
Horowitz, N.H., 220
Hot springs, 97
Hydrogen, molecular, 74–75, 78–79, 87
Hydrogen ions, 102–104
Hydrogen sulfide, 91–94, 107–109, 126, 138
Hydrothermal vents, 107

Immunity, 157–168, 173
Infectious disease, 144–161, 166, 183
Influenza virus, 177? Inorganic compounds, 24
Insulin, 202
Intestinal bacteria, 46, 73, 167
Iron, use by bacteria, 105

Jacobsen, J.C., 18
Jenner, E., 159

King George I, 159
Koch's Postulates, 157
Koch, R., 35, 37, 39–40, 155, 157
Küster, E., 171–172
Kützing, F., 16

Lactic acid, 14–15
Lactobacillus arabinosus, 133
Lactobacillus bulgaricus, 52
Lactobacillus sanfrancisco, 216
Landfills, 75–76
Leaching of metal ores, 94
Leeuwenhoek, A.v., 2–8, 57, 212
Legionella pneumophila, 216
Leprosy, 145, 157
Leuconostoc mesenteroides,216
Levi, P., 70
Lewis, G.N., x Lichens, 115
Liebig, J., 17
Lipids, 30–31
Lipman, C.B., 57
Lockjaw, 56
Longevity of microbes, 56, 58–60
LSD (lysergic acid diethylamide), 115
Luria, S.E., 196, 202
Lyophilization, 51
Lysozyme, 163

Magnetite, 104–106
Magnetotactic bacteria, 104–105
Malt, 26
Marsh gas, 72–73
Martian monsters, 217
McCarty, M., 204
Measles, 176
Meiklejohn, J., 84
Meister, J., 156
Membranes, 31
Mesophiles, 97
Metabolism, 28–29, 31
Metabolite, defined, 182
Metal corrosion, 92
Metcalf, L., 137

Metchnikoff, E., 163–164
Methane, 24, 65, 77
Methane, formation of, 72–78, 109
Methane, hazards of, 75
Methane, utilization of, 79–81
Methanobacterium, 43
Methanobacterium thermoautotrophicum, 65
Methanococcus voltae, 73
Methanogens, 73–75, 78, 109, 141, 211
Methyl alcohol, 79–81
Methylotrophs, 79–81
Microbe-animal associations, 106
Microbes, discovery of, 2
Microbiological assay, 134
Micrometer, defined, 3
Microscopes, 2–4
Mildewed fabrics, 113
Mineralization, 65, 138
Molds, 113
Molecule, definition of, 21
Montagu, Lady Mary, 158–159
Morowitz, H.J., 9
Mucor hiemalis, 216
Multiplication, cell, 39
Mummy grain, 58
Muscardine, 149–150, 152
Muscle cells, 15
Mutation, 195–196
Mycobacterium leprae, 157
Mycobacterium tuberculosis, 43
Mycoplasma genitalium, 216
Mycoplasma hypopneumonia,216
Myxobacteria, 110–112

Napoleon, 72, 84, 146
Needham, J., 25
Neisseria gonorrhoeae, 177
Niacin, 130–134
Niacinamide, 132–133
Nitrate, 82–84, 138, 141
Nitrate assimilation, 83
Nitrification, 83–84
Nitrogen cycle, 82–90
Nitrogen fertilizer, 87
Nitrogen fixation, 36, 52, 65, 85–90, 125–126
Nitrogenase, 86, 88
Nitrosomonas europaea, 65, 82
Nostoc, 125
Nostoc muscorum, 65
Nucleic acid bases, 189–191, 194–196
Nucleic acids, 25, 31, 44, 48, 80, 179
Nucleus, cell, 10

Oil, 70–71
Organic compounds, 14, 24
Origin of life, 208
Oscillatoria, 125
Osmophiles, 98–99
Osmotic pressure, 99
Oxygen, 14, 17, 47–48, 66, 69, 75, 85, 88–89, 105, 107–108, 124–125, 138, 141, 208–209

Paecilomyces carneus, 216
Paik, G., 59
Pandemics, 146
Panspermia, 208
Papain, 30
Paratyphoid fever, 162
Parduhn, N.L., 60
Pasteur Institute, 156
Pasteurization, 14, 181
Pasteur, L., 1, 13, 17–18, 34–35, 58, 148, 154–157, 165
Patented microbes, 206
Pathogenic microbes, 40, 141, 144, 181
Peat, 70
Pellagra, 132–133
Peloponnesian War, Great, 145
Penicillin, 162, 185–186, 204
Penicillium, 113–114
Penicillium chrysogenum, 186
Penicillium notatum, 184, 186
Pentagon papers, hydropulped, 114
Pentagon papers, sugar from, 114

Petri dish, 40, 42
Petri, R.J., 40
pH scale, 102–104
Phagocytes, 163–164
Phipps, J., 159
Phormidium frigidum, 98
Photoautotrophs, 125–126
Photoheterotrophs, 126
Photosynthesis, 63, 65–70, 108, 122–123, 126–127, 209
Photosynthetic bacteria, 32, 48, 88, 93, 98, 104, 124–125, 127
Piranesi, G., 136
Plagues, 144–148, 150, 152
Plant seeds, 57–58
Plasmids, 192, 199–201
Poliomyelitis, 176
Poliomyelitis virus, 165, 177
Polio vaccine, 165
Polysaccharide, 26
Postgate, J., 90, 93, 138
Priestley, J., 66–68
Procaryotes, 10
Protein food supplements, 80
Proteins, 22, 24–25, 27–29, 31, 44, 48, 91, 193–196, 209
Protozoa, 10, 74, 78
Prout, W., 24
Pruteen, 81
Psychrophiles, 98
Pure cultures, 36–37, 39, 41–42
Purple bacteria, 124–127
Pyrite, 94

Quayle, R., 74

Rabies, 156, 165
Rahn, O., 63, 142
Randall, M.,
Recombinant DNA, 205
Restriction enzymes, 199
Reyniers, J., 171
Rhizobium, 89
Rhizopus oryzae, 223–224, 227
Rhodopseudomonas, 125
Rhodospirillum, 43, 125
Rhodospirillum rubrum, 177
Ribonuclease, 29
Rice paddy fields, 126
RNA, 25, 179, 209–211
Rosebury, T., 166
Rosenburg, F., 195
Rotifers, 57
Rumen symbiosis, 73–75

Saccharomyces bailii,var. osmophilus,216
Salk polio vaccine, 165
Salmonella typhi, 52
Salt pans, 99
Saltpeter, Chile, 84
Sauerkraut, 203
Schwann, T., 16
Seeds, plant, 57–58
Sewage, 73, 76, 135–142
Seward, M.R.D., 59
Sexuality in bacteria, 187
Shapes of bacteria, 43
Shaw, G.B., 183
Silkworms, 149–152, 155
Single-cell protein, 81
Skin, microbes on, 169–170
Smallpox, 158–160, 176
Smith, E.F., 155
Sneath, P.H.A., 58
Soil, 10–11, 85
Solar energy, 69, 71
Spiroplasma citri, 216
Spontaneous generation, 5, 33–34, 153–154
Spores, 113–114
Stanley, W.M., 176
Staphylococci, 183–184
Starch, 26–27, 123
Sterilization, 56, 181
Stigmatella aurantiaca, 111–112
Strains of microbes, 46
Streak plate (dish), 41
Streak slide, 38
Streptococcus, 43
Streptococcus lactis, 43
Streptomyces griseus, 52
Streptomycin, 185

Sucrose, 26
Sugar, 14, 16–17, 24–26, 69, 122–123
Sulfa drugs, 183
Sulfate, 91–93, 109, 126, 138
Sulfur cycle, 91–94
Sulfur dioxide, 91
Sulfuric acid, 94
Swift, Jonathan, 121
Synechococcus, 125
Synthetic media, 45, 47
Syphilis, 148, 152

Taste buds, 6
Teich, M., 24
Termites, 77–78
Thermoactinomyces, 53, 59–60
Thermoanaerobacter ethanolicus, 97
Thermophiles, 97, 211
Thermoplasma acidophilum, 104
Thermus thermophilus, 97
Thiobacillus concretivorous, 94
Thiobacillus ferrooxidans, 94, 103, 216
Thiobacillus thiooxidans, 65, 94, 216
Thucydides, 145
Tobacco mosaic virus, 175–176
Torula, 225
Toxins, 162
Traveller's diarrhea, 46
Trichoderma viride, 114
Tube worms, 107–108
Tuberculosis, 37, 43, 157
Tyndall, J., 154
Typhoid, 136, 162, 183
Typhus, 148

Ultraviolet radiation, 182
Unsworth, B.A., 59
Updike, J., 71
Uranium mining, 94

Vaccination, 158–160, 165
Vampirococcus, 127
Viking Mars probe, 219–220
Vindolanda, archeological site, 59
Viruses, 160, 175–180, 186
Vitamins, 25, 129–134, 173, 224–225
Volcani, B., 100
Volta, A., 72–73

War of the Worlds, 217
Watson, J., 192
Watterson, J.R., 60
Wedgewood, J., 68
Welles, O., 219
Wells, H.G., 217, 219
White blood cells, 162, 165–166, 174
Wiley, A.J., 57
Wine, 16, 18, 70
Witchcraft trials, 115
Wöhler, F., 17
Wright, A., 183

Yarrow, P.J., 149
Yeast extract, 45
Yeasts, 10, 14, 16–17, 27, 31, 80, 113, 118–120, 199, 222, 224–226
Yellow fever, 176
Yersin, A., 152
Yersinia pestis, 152
Young, P., 98

Zilinskas, R.A., 206
Ziman, J.M., ix Zimmerman, B.K., 206
Zinsser, H., 146
Zymomonas mobilis, 52